Enterprise AI Observability and Monitoring

Monitoring, Governing Production AI Systems
Drift Detection, LLM Monitoring, Agentic AI, Governance, and FinOps for Production

First Edition

Jordan Louis-Charles

Cybersoft Publishing LLC
Fort Washington, MD 20744
Drclaude.net

First Edition, April 2026

Table of Contents

1 Introduction

Production AI systems are making predictions right now. Some of those predictions are wrong, and no alert has fired. The model server is healthy. Latency is within SLO. The inference endpoint returns a 200. But the fraud model trained on pre-pandemic behavioral signals is scoring transactions against a distribution that no longer exists. The loan model's upstream feature pipeline silently changed its encoding three months after deployment. The LLM-powered support assistant started hallucinating refund policy details after a prompt template was promoted without a regression eval. Every one of these is a live production incident. None of them looks like one in your existing dashboards.

This book is written for the engineers and architects who actually build, install, and run these systems. If you are a staff or principal ML engineer, an ML platform engineer, an SRE carrying on-call for inference infrastructure, a technical architect designing an AI observability stack, or a security engineer thinking through audit log immutability for agentic systems, this is your working reference. It is not a strategy deck. It is a playbook: concrete instrumentation patterns, failure taxonomies with root-cause analysis approaches, SLO design for non-deterministic systems, governance as an artifact pipeline problem, and FinOps as a unit-cost telemetry problem.

The five-layer AI observability stack: infrastructure, data pipeline, model behavior, output quality, and business outcome layers, showing where each class of silent failure originates

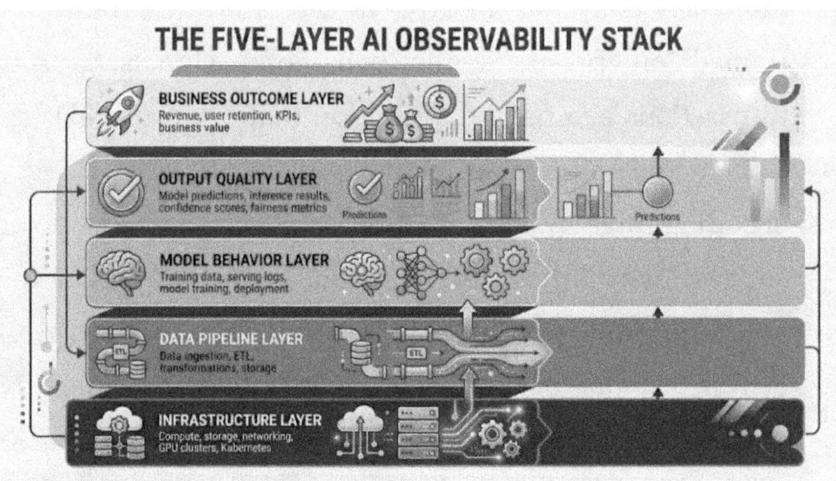

Traditional DevOps observability was built for deterministic systems. A service is up or down. A response is correct or an exception. RED and USE method metrics that work perfectly for a microservice do not capture the failure modes that matter for a production ML model. A model server can be fully available under a latency SLO, with zero 5xx errors, and silently be wrong on 30% of inputs because the input feature distribution shifted outside the training envelope. You cannot detect that with Prometheus scrape intervals and uptime checks. You need a second observability plane layered above infrastructure, one that speaks in PSI scores, KS statistics, embedding drift vectors, ROUGE deltas, and tool-call error rates.

The three major AI system classes each introduce distinct instrumentation requirements. Classical predictive ML models fail due to data drift, concept drift, feature pipeline mutations, and training-serving skew. Large language models introduce hallucination rates, semantic drift in retrieved context, prompt-injection surface, and output-schema violations. Autonomous agentic systems compound these with multi-step execution failures, tool-call error chains, runaway cost loops, and emergent behaviors that no single span captures. A unified observability architecture must accommodate all three classes on a shared telemetry backend and alerting pipeline.

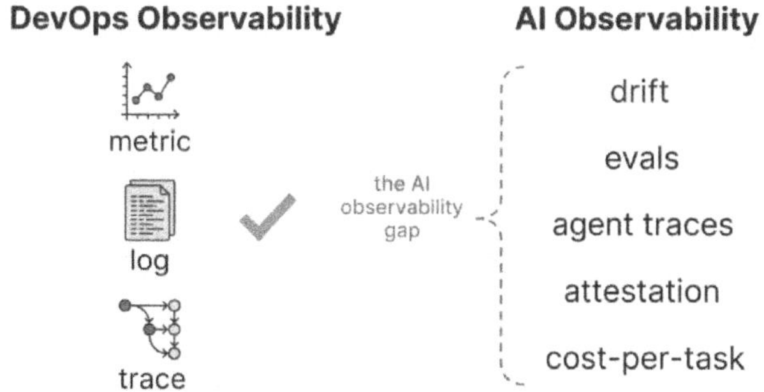

The AI-vs-DevOps observability gap: classical APM covers infrastructure and service health, while AI observability must additionally cover output quality, drift, and behavioral correctness across the full model lifecycle

The book is organized across fifteen chapters in four arcs. The first arc (Chapters 1 through 4) lays the foundation: the failure taxonomy, the five-layer observability stack, OpenTelemetry-native instrumentation patterns, and drift-detection strategies (PSI thresholds, evaluator-as-judge pipelines, span-linked tool-call audits) that make silent failures visible. The second arc (Chapters 5 through 7) covers monitoring by system class: reference windows and shadow deployment for classical ML, prompt telemetry and retrieval-quality signals for LLMs, and agent-loop instrumentation with cost-sentinel design and kill-switch architecture. The third arc (Chapters 8 through 10) covers the operational layer: vendor evaluation, data architecture (schema registry, feature store integration, immutable audit logs), and SLO design for AI-specific SLIs with MTTD and MTTR improvement patterns. The fourth arc (Chapters 11 through 15) covers governance as an artifact pipeline and lineage graph problem, FinOps as unit-cost token accounting with budget guardrail instrumentation, organizational structure, SR 11-7, and EU AI Act compliance, mapped to concrete engineering artifacts, and the forward-looking architecture questions the discipline is moving toward.

Every chapter ends with a Technical Team Checklist and Team Conversation Starters. Each checklist item names a concrete action, assigns an owner role (ML Platform Eng, SRE, ML Engineer, Data Eng, Security Eng, Governance Eng,

FinOps Eng, or Architect), and states a measurable done-when criterion. These slot into sprint planning or post-incident reviews. The conversation starters are questions to bring into design reviews or retrospectives when the chapter's topic is live on your team.

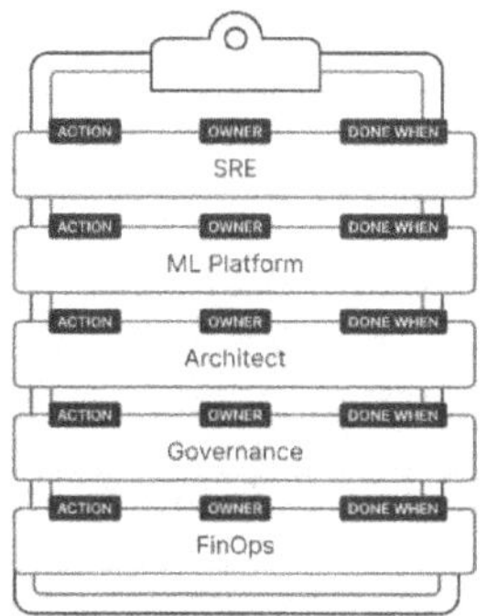

Structure of the Technical Team Checklist used at the end of every chapter: action, owner role, and measurable acceptance criterion, designed for sprint planning and post-incident retrospective use

Before Chapter 1, a few terms need precise definitions. Observability means having enough structured telemetry to diagnose any failure mode, including ones not anticipated when writing the monitors. Monitoring is a subset: it fires when a known metric crosses a known threshold. The most consequential AI failure modes never cross a threshold configured in advance. Data drift is a distributional shift in raw input features. Concept drift is a change in the mapping between features and target outcomes. Feature drift is a change in engineered features caused by pipeline modifications or upstream schema changes. Model decay is the cumulative decline in quality when these mechanisms go unaddressed. A trace, in the OpenTelemetry sense, is the directed acyclic graph of spans that represents a request's path through a distributed system, and it is the primary structure for root-cause analysis of multi-step agentic failures.

Vendor callouts are illustrative rather than prescriptive: a tool like Arize Phoenix, Langfuse, WhyLabs, or Fiddler may implement a given pattern; the pattern is what the chapter is teaching. The goal is for your team to instrument, monitor, and govern a production AI system regardless of which tools you have standardized on, and evaluate any new tool against your platform's actual requirements.

The AI systems your organization has shipped are live infrastructure. They serve users, affect decisions, accumulate behavior under distribution shift, and carry compliance exposure. They deserve the same operational discipline applied to any other production system, with instrumentation and detection strategies tuned to the failure modes that actually occur in AI. This book provides those strategies in a form that a working engineering team can apply directly.

2 The Hidden Blind Spot in Your Enterprise AI Strategy

2.1 Opening Scenario

The pager fires at 3:14 AM on a Tuesday. The on-call rotation lands on the staff ML engineer who owns the revenue-attribution pipeline for the company's primary e-commerce recommendation model. Every infrastructure metric is green: p99 latency at 42ms, error rate at 0.03%, GPU utilization at 61%. The alert is from a feature-store data quality monitor, a senior data engineer wired up six weeks ago as a post-incident action item. PSI on the user-recency feature: 0.31, past the 0.2 warning threshold.

The revenue dashboard has been diverging from the model's predicted conversion signal for 19 days. Day one: 2% gap, attributed by finance to a promotional skew. Day eleven: 7%. Tonight: 14%. No model-side alert ever fired because the model server has no awareness of output distribution. It returns valid JSON at normal latency and logs nothing about whether those predictions are coherent with its training distribution.

Root cause surfaces in 22 minutes: an upstream data vendor changed the schema of a session-duration field on day one of the divergence window, capping values at 3600 seconds instead of passing raw milliseconds. The feature engineering job ingested the capped values without complaint. The model consumed them without complaint. The revenue signal complained, but nobody had wired that signal to a model health monitor.

Total estimated revenue impact from 19 days of degraded recommendations: $380,000. MTTD: 19 days. MTTR once the alert fired: 38 minutes. The ratio of silent damage to active remediation is approximately 730-to-1. That ratio is the argument for AI observability.

2.2 Why It Matters for Production Systems

The hard part of production AI is not the model. It maintains a trustworthy signal about what the model is doing after the deployment pipeline finishes. A model can be perfectly healthy by every infrastructure SLI and simultaneously produce degraded predictions that quietly erode a business process. A model is a function optimized for a training distribution. The moment it enters production, the world begins changing. User behavior shifts. Vendors change schemas. Seasonal patterns arrive that were underrepresented in the training window. None of these changes generates an exception or causes a 5xx response. The failure mode is semantic, not syntactic, and infrastructure-layer tooling has no vocabulary for it.

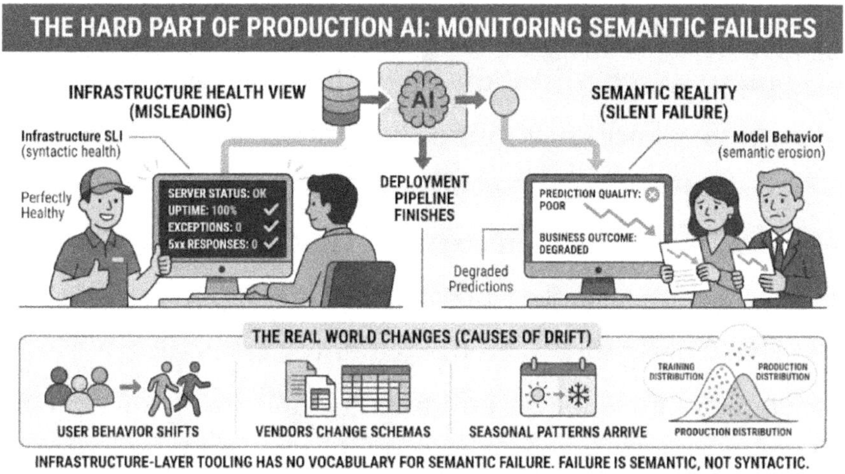

INFRASTRUCTURE-LAYER TOOLING HAS NO VOCABULARY FOR SEMANTIC FAILURE. FAILURE IS SEMANTIC, NOT SYNTACTIC.

In a traditional service, behavioral degradation generates a signal: elevated error rates, increased latency percentiles, failing health checks. In a drifted ML system, the signal lives outside the model server. It lives in the statistical relationship between today's input distribution and the reference window from training, in the divergence between predicted outcomes and observed ground-truth labels, and in downstream business metrics that may not be instrumented at a granularity that makes attribution to the model obvious. Without monitoring output distribution, a model can drift for weeks before a business analyst notices a KPI anomaly. MTTD in organizations without automated model monitoring ranges from weeks to months. LLMs introduce additional failure modes: hallucination-rate drift, output-length distribution shifts, toxicity emergence, and reasoning-quality degradation, with no crisp statistical signal. Wiring an LLM endpoint requires span-linked traces that capture the prompt template version, retrieved context, output token distribution, evaluator-as-judge scores, and user feedback in a unified telemetry stream. Purpose-built platforms (tools like Arize, WhyLabs, or Fiddler) and emerging OpenTelemetry conventions for AI-specific spans make this tractable. Still, the engineering investment is non-trivial: designing reference windows, calibrating PSI and KS-statistic thresholds, building ground-truth join pipelines, and integrating drift alerts into the on-call rotation.

2.3 From Pilot to Production: Closing the Observability Gap Before Models Go Dark

A successful pilot is a controlled experiment: curated data, a representative holdout set, and close supervision from the data science team. That environment shares almost nothing with the production environment the model will occupy for the next 18 months. Production data changes every day, and production pipelines have upstream schema dependencies that can mutate silently. The pilot-to-production gap is not a model capability gap. It is an observability infrastructure gap. Because that infrastructure does not exist at pilot time, it is typically the last thing built before launch, built incompletely, or not built at all.

Why Production AI Models Fail Silently After Testing

The silence of model failure is its defining characteristic. A conventional software defect generates observable artifacts: a stack trace, an HTTP 500 response, or a failed assertion. A model that has drifted generates none of these. It produces predictions: syntactically valid, correct data types, within expected value ranges, returned within SLO. The only thing wrong with them is that they are no longer accurate, and accuracy is not a property of infrastructure monitoring. For a tabular classification model, the failure is typically feature drift: PSI and KS-statistic tests over a rolling reference window detect this, but only if those tests are running and their alerts are wired into the on-call rotation. For an LLM, failure modes are broader: a prompt template subtly edited to alter output quality, a retrieval index that grows stale, or a provider model update that shifts behavior without notice. In each case, the failure is invisible to the inference endpoint. The observability layer must exist outside the model server, monitoring the statistical properties of inputs and outputs over time and firing alerts when divergence exceeds thresholds calibrated by business impact analysis.

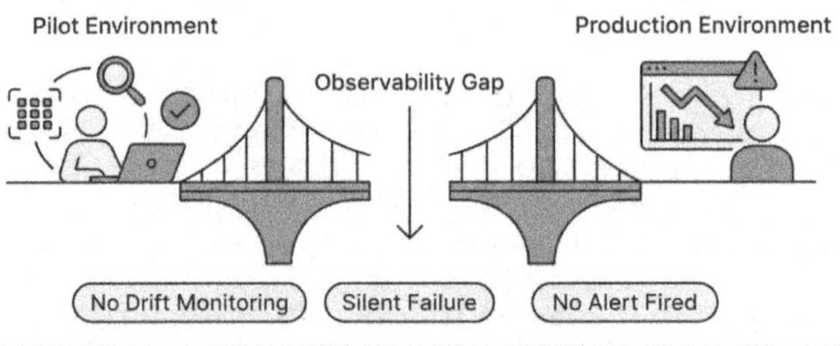

PSI and KS-statistic trends over a 30-day reference window showing distribution divergence between pilot validation data and live production traffic for a tabular classification model

What AI Observability Actually Means for Production Systems

For a production AI system, the three pillars of metrics, logs, and traces each require ML-specific content that standard APM instrumentation does not produce. Metrics must include output distribution statistics: mean prediction score, score histogram, PSI against the reference window, and KS-statistic for continuous features. Logs must include prediction records linked to ground-truth outcomes. A prediction log that is never joined to a ground-truth outcome cannot support performance evaluation. The data engineering work to build that join pipeline is consistently the most underestimated part of an AI observability buildout. Without it, the monitoring layer can detect distribution drift but cannot measure actual accuracy degradation. Traces for LLM or agentic systems must span the full inference stack: prompt construction, retrieval pipeline execution, LLM API call, and output parsing. Span linking across the retrieval pipeline and the generation call, following OpenTelemetry semantic conventions for LLM spans, makes an LLM system debuggable when quality regresses.

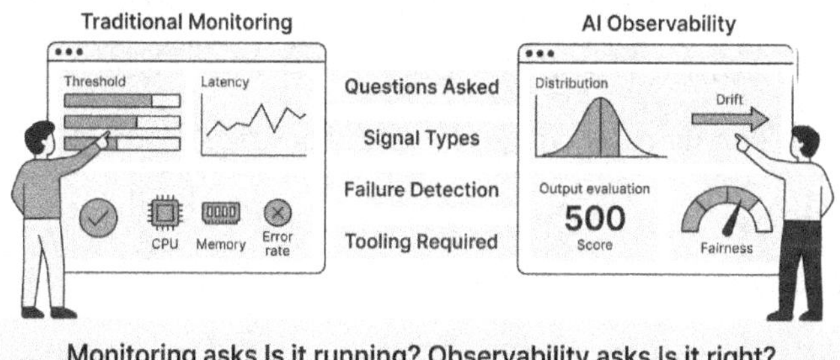

Three-pillar AI observability architecture: output distribution metrics pipeline, prediction log store with ground-truth join, and distributed trace with span linking across retrieval and generation stages

Three Structural Conditions That Create AI Monitoring Blind Spots

Blind spots in AI production are structural, not accidental. Three conditions reliably produce unobserved AI systems. The first is ownership ambiguity: data science, data engineering, the ML platform team, and SRE each own a slice of the stack, but no single team owns the cross-cutting observability layer that spans all four domains, and each assumes the others have it covered. The ML platform team is the natural owner of this responsibility, but it requires an explicit charter and budget established before models reach production. The second condition is stage-gate thinking: treating model deployment as a project completion event leads to resource reallocation immediately after launch, leaving the model without a dedicated monitoring cadence, a defined retraining trigger, or a named owner for operational health. The third condition is metric mismatch: latency and error rate are not the metrics that matter for model health. The ML platform team must build additional instrumentation to produce PSI, the output score distribution, feature importance drift, ground-truth accuracy, and business KPI correlations. Building these before launch as a required gate is far less expensive than retrofitting them afterward.

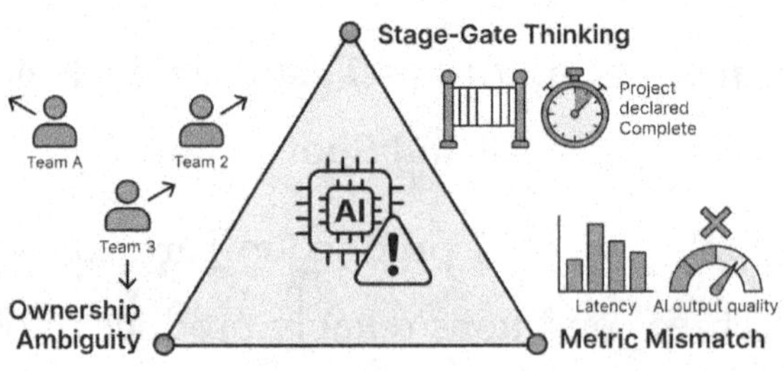

Ownership responsibility map for AI production observability: the four-team coverage gap across feature pipeline, model serving, output monitoring, and ground-truth feedback, with the ML platform team as the cross-cutting owner

2.4 The Real Cost of Letting Production Models Run Unmonitored

Unobserved AI systems erode quietly, and costs accumulate across three dimensions: operational efficiency losses from degraded model quality, compliance exposure from audit-trail gaps, and reputational damage from visible failure events. Each is quantifiable and maps to risk categories that engineering teams already own.

Revenue Leakage from Silent AI Model Drift and Degradation

Every production model drives or informs a business process. When these models drift, the downstream process operates on systematically wrong information. The revenue impact is the difference between what a correctly calibrated model would have recommended and what the drifted model is recommending, multiplied by decision volume and dollar value per decision. Revenue leakage rarely manifests as a discrete event. It appears as a slight decline in conversion rate, a modest increase in return rates, and a gradual uptick in churn that analytics attributes to seasonal variation. Without a monitoring layer correlating business metrics with model health signals, organizations absorb losses for quarters before tracing them back to the model. Quantifying revenue sensitivity before deployment is the engineering foundation for alert threshold design: estimate the business outcome per prediction, the sensitivity to a 10-percentage-point degradation in accuracy, and the current MTTD for a drift event of that magnitude. A PSI threshold calibrated to model-specific revenue impact is defensible engineering. A uniform 0.2 threshold applied regardless of impact is a guess.

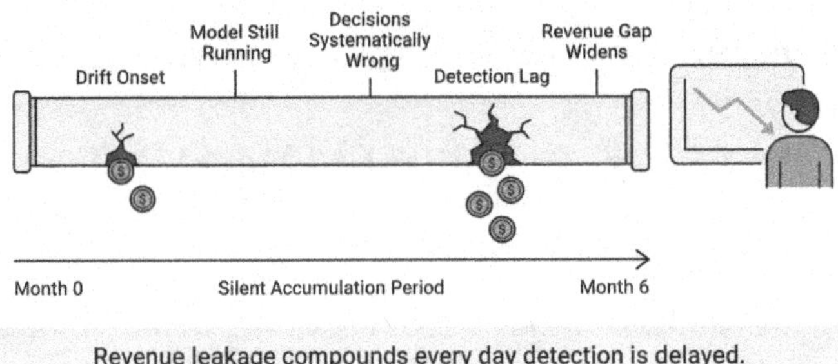

Revenue sensitivity model: daily business outcome loss as a function of accuracy degradation percentage, used to back-calculate PSI and KS-statistic alert thresholds for a production recommendation model

AI Compliance Exposure and Regulatory Risk from Model Drift

Regulatory frameworks governing AI systems share a common requirement: not just that a model was validated before deployment, but that it was monitored continuously in production. SR 11-7 requires banks to maintain evidence of ongoing monitoring of model performance. The EU AI Act requires high-risk AI systems to include logging and monitoring mechanisms to detect serious incidents. HIPAA enforcement has extended to AI-assisted diagnostic tools, requiring ongoing monitoring of bias and accuracy. Every regulated production AI system needs an immutable audit log of monitoring events: drift metric values at regular intervals, alert firings and their disposition, retraining events with before-and-after performance comparisons, and manual interventions. Compliance exposure compounds when model degradation has a disparate-impact dimension: a credit-scoring model that drifts in a way that disadvantages a protected class creates fair-lending liability difficult to defend without a continuous monitoring record. An observability layer that tracks output score distributions by demographic segment and alerts when gaps widen provides both the detection capability and the audit evidence a defensible compliance posture requires.

Compliance Risk from Unobserved AI

Regulators ask whether monitoring existed – not just whether the model passed.

Immutable audit log architecture for a regulated ML system: monitoring event stream, alert disposition records, and retraining attestation in an append-only store with cryptographic hash chaining for tamper evidence

AI Reputational Damage and the Customer Trust Erosion Cycle

The public narrative for an AI failure is almost always about whether the organization was paying attention. A bank whose credit model drifted for six months and harmed customers faces a different framing than one whose model experienced a detected, promptly remediated drift event. The difference is not the model. It is the monitoring infrastructure. The erosion of trust follows a predictable pattern: a model degrades, a user or regulator encounters an anomalous output and investigates, and the organization cannot produce a monitoring record. The resulting narrative has a blast radius disproportionate to the technical incident because it speaks to organizational character. When an incident occurs, the ML platform team must be able to produce: a timeline of monitoring metric values leading up to the incident, the alert that fired or an explanation of why it did not, the root cause analysis grounded in telemetry data, and the corrective action plan with acceptance criteria. That incident response package separates a contained technical incident from a reputational crisis.

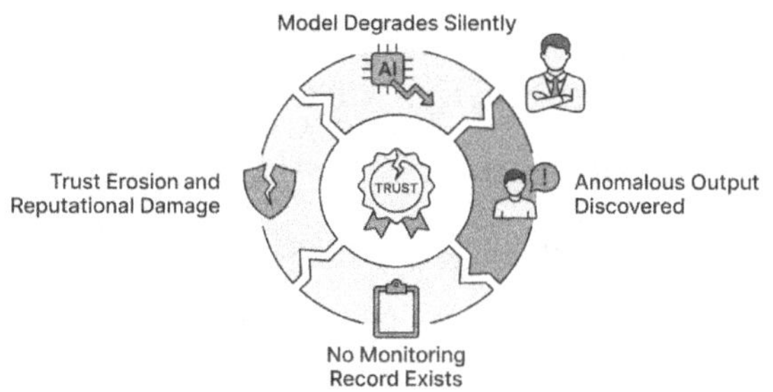

Reputational risk timeline comparing MTTD and blast radius for monitored versus unmonitored production AI systems across a 90-day incident window

2.5 Building the Business Case for AI Observability Investment

An AI observability program requires investment in budget, staffing, and tooling. The case rests on three pillars: quantified risk reduction from shorter MTTD and MTTR, demonstrated return from industry data, and the compounding value of shared observability infrastructure across the portfolio.

AI Observability 357 Percent ROI: What the Research Measures

Forrester Research's Total Economic Impact study on AI observability platforms found that organizations with mature monitoring programs realized a 357 percent return on investment over three years, accounting for avoided incident costs, reduced MTTR for model failures, labor savings from automated drift detection, and value recaptured from prematurely retired models. The largest component is avoided incident cost: the study found that in organizations without automated model monitoring, the average gap between drift onset and detection was 73 days. Automated monitoring reduces MTTD from 73 days to 2-5 days for statistically detectable distribution-drift events. The second component is MTTR reduction: when a drift event fires an alert with a statistical context attached, root-cause investigation starts with data rather than exploratory log searches. The third component is to avoid unnecessary model retirement. Without output distribution monitoring, the team cannot distinguish "this model has drifted and needs retraining" from "this feature pipeline has a quality regression and the model is fine." The cost of an unnecessary retraining cycle is often between $30,000 and $150,000.

Avoiding two or three unnecessary retraining cycles per year more than pays for a monitoring platform.

AI Observability ROI Value Drivers

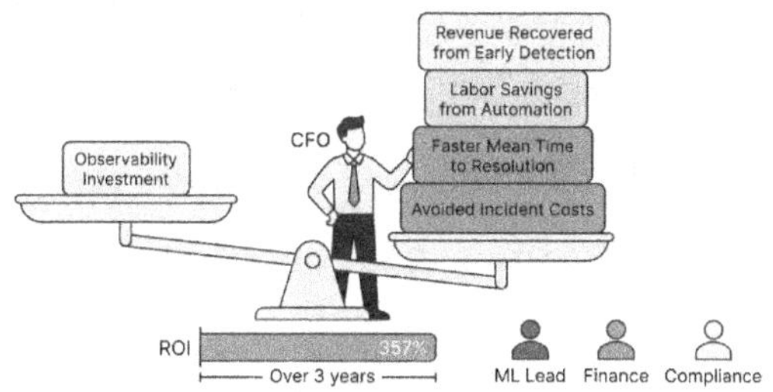

The ROI of observability is the cost of undetected degradation prevented.

Three-component ROI decomposition for AI observability investment: MTTD reduction value, MTTR reduction value, and avoided unnecessary model retraining cost, aggregated across a five-model production portfolio.

AI Observability as Shared Portfolio Infrastructure for MLOps

The correct accounting unit for AI observability investment is the model portfolio, not the individual model. Observability infrastructure has high fixed costs and low marginal costs: the first instrumented model requires a complete platform buildout, whereas subsequent models inherit that infrastructure and require only model-specific configuration. By the time the platform covers ten models, the per-model cost of adding monitoring is a small fraction of the initial investment, while aggregate risk coverage has grown tenfold. Building observability after models are already running requires retrofitting instrumentation into systems not designed to emit the required data. Building observability-ready systems from the start, using standardized telemetry with OpenTelemetry conventions, is far less expensive. The ML platform team's role is to establish observability as a foundational platform service and enforce it through a production-readiness review gate that requires observability artifacts before any model is promoted.

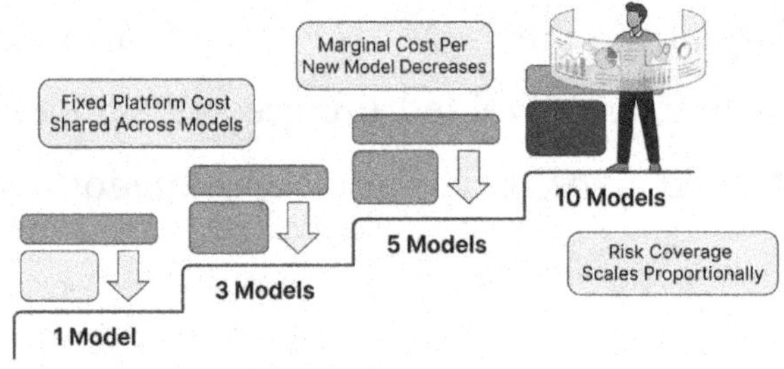

Observability infrastructure becomes cheaper per model as the portfolio grows.

Per-model marginal cost curve for AI observability as portfolio size grows, showing fixed-cost platform amortization against risk coverage growth, with cost per model converging toward near-zero at scale

How to Build an Internal AI Observability Funding Argument

The internal funding argument for an AI observability program must be framed as a risk inventory rather than a capability roadmap. For each production model, record: the business process it drives, the estimated daily revenue or risk exposure if the model degrades by 10%, the current MTTD for a drift event of that magnitude, and current monitoring coverage. The product of daily exposure and MTTD gives the per-incident expected loss. In most enterprise AI portfolios with five or more production models, the aggregate exposure figure, compared to the all-in cost of an observability program (engineering time, tooling licenses for a tool like Arize, Langfuse, or WhyLabs, and operational overhead), produces a compelling positive result before accounting for compliance risk avoidance. The funding argument should be presented as a phased roadmap: Phase one instruments the highest-risk production models with full observability within 60 days; Phase two establishes shared platform infrastructure and brings all current production models to a monitoring baseline within six months; Phase three implements advanced capabilities (LLM-specific monitoring, evaluator-as-judge pipelines, agentic trace linking, compliance audit logging) as the portfolio grows. The alternative to investment is not zero cost. It is the unbounded, unquantified cost of operating AI systems that go unwatched.

Internal AI Observability Funding Argument

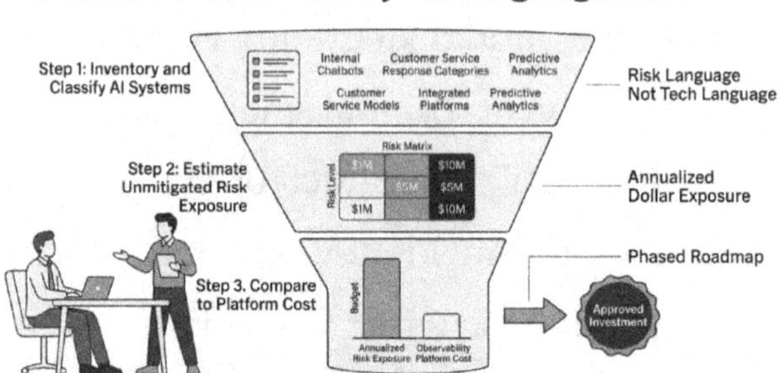

Frame observability as risk mitigation, not infrastructure spending.

Three-phase AI observability buildout roadmap: Phase 1 high-risk model instrumentation at 60 days, Phase 2 shared platform baseline at 6 months, Phase 3 advanced capabilities at 18 months, with estimated portfolio risk coverage at each phase

2.6 Technical Team Checklist: Pre-Production Observability Requirements

- **Action:** Compute estimated daily revenue or risk exposure per 10% accuracy degradation for each production model and record it in the model card before production promotion. **Owner:** ML Eng. **Done when:** Every model in the model registry has a documented sensitivity estimate and a recommended alert threshold.
- **Action:** Instrument the model serving layer to emit prediction inputs and outputs to a prediction log store at a configurable sampling rate, with a unique inference ID that supports downstream ground-truth joins. **Owner:**

ML Platform Eng. **Done when:** Prediction logs are queryable in the central data warehouse within 15 minutes of inference, with inference ID as a joinable key, for every production model.

- **Action:** Build and validate a ground-truth join pipeline linking prediction log records to downstream outcome labels or transaction records, with a defined join latency SLO appropriate to the business process. **Owner:** Data Eng. **Done when:** Ground-truth accuracy rate is computable on a 24-hour lag or better for every model class where labels are available within that window.
- **Action:** Configure PSI and KS-statistic monitoring over a 30-day reference window for all monitored input features, with alert thresholds calibrated from revenue sensitivity estimates, and integrate alert routing into the existing on-call rotation. **Owner:** ML Platform Eng. **Done when:** PSI and KS-statistic trends are visible in the same dashboard as latency and error rate, and a test alert has been fired, routed, and acknowledged by the on-call rotation.
- **Action:** Author a model incident runbook covering drift detection response, root cause investigation steps, rollback procedure, and retraining trigger criteria for each production model. **Owner:** ML Eng. **Done when:** The runbook has been reviewed in a production readiness review, linked from the on-call alert, and

walked through in at least one tabletop exercise before the model goes live.

- **Action:** Establish a production readiness review gate requiring observability artifacts (prediction log schema, monitoring dashboard, alert configuration, runbook link) as non-negotiable criteria before any model is promoted to production. **Owner:** Architect. **Done when:** The production readiness checklist includes observability as a gate item, and at least one model has been held from promotion until observability is complete.
- **Action:** Require all new model serving deployments to emit standardized telemetry using OpenTelemetry semantic conventions, including model name, model version, prediction latency, and output score as span attributes. **Owner:** ML Platform Eng. **Done when:** OpenTelemetry instrumentation is present in the model serving base image and verified by the CI pipeline for all new deployments.
- **Action:** Define a maximum tolerable detection window (MTDW) for each production model class, document it as an operational SLI, and compute the current MTTD baseline from historical incident data. **Owner:** SRE / On-Call. **Done when:** MTDW and MTTD baselines are documented in the service reliability record for each production model, with a gap analysis identifying models outside acceptable tolerance.

- **Action:** Implement immutable audit logging for all monitoring events, alert firings, and model interventions in a compliance-ready append-only store with a documented retention policy. **Owner:** Governance Eng. **Done when:** Audit log is queryable, tamper-evident, and covers the full event history for all production models, with a documented response procedure for regulatory audit requests.
- **Action:** Build a shared observability cost model, amortizing platform infrastructure across all production models and reporting per-model monitoring cost to the owning engineering team monthly. **Owner:** FinOps Eng. **Done when:** Per-model observability cost is visible in the engineering cost dashboard alongside compute and storage costs.

2.7 Team Conversation Starters

- When our last production model was deployed, what was the MTTD for a drift event at that time, and what evidence do we have that it has improved since?
- For each current production model, do we have a ground-truth join pipeline running, and if not, how are we planning to measure accuracy degradation in production rather than only inferring it from distribution shift?

- If the most business-critical model we run today started drifting tonight, what would be the first signal anyone would see, which team would see it, and what runbook would they follow?
- Are our PSI and feature drift alert thresholds derived from revenue sensitivity analysis for each specific model, or set to a generic default with no relationship to the actual business impact of that model?
- Does our production readiness review treat observability as a gate item with defined acceptance criteria, or as a recommendation that teams can defer after launch?
- What is the aggregate annualized risk exposure across our unmonitored or partially monitored production models, and does that number appear anywhere in engineering planning conversations?

2.8 Key Takeaway

The central operational risk in enterprise AI is not that teams build bad models. It is when teams deploy good models into environments where no one is watching that what happens next. Model drift, upstream data schema changes, and distribution shift are the normal operating conditions of any production system that runs long enough. The engineering discipline of AI observability, covering output distribution monitoring, ground-truth feedback pipelines, statistical drift detection wired into on-call alerting, and immutable audit logging, converts those normal operating conditions from invisible risk into manageable incidents with defined MTTD, MTTR, and blast-radius bounds. Teams that instrument this layer before their first model ships operate with radically shorter detection windows, smaller incident blast radius, and audit evidence trails that satisfy regulatory scrutiny. Teams that defer it are accumulating unbounded, silent exposure.

3 From DevOps Observability to Production AI Observability

3.1 Opening Scenario

It is 2:17 AM on a Tuesday. The on-call SRE for a mid-sized e-commerce platform gets paged: order processing latency has spiked to 4.2 seconds, double the SLO threshold of 2.0 seconds. She pulls the runbook, opens Grafana, and works the usual checklist. Kubernetes pods are healthy. Database connection pool utilization sits at 34 percent. The LLM gateway service handling SKU-to-description expansion reports nominal infrastructure health: p50 at 280 milliseconds, p99 at 810 milliseconds, zero HTTP 5xx errors. Every metric is green.

She pages the ML platform engineer on secondary rotation. He joins the bridge and pulls raw completions from the LLM gateway logs. The pattern is immediately obvious and deeply wrong. The gateway is generating SKU codes that do not exist in the catalog. A prompt that should return "SKU-88821-BLK-L" is returning "SKU-88821-BLK-XL-PRIME," a code that has never existed. The model is hallucinating catalog identifiers, and the downstream order service is failing lookup, retrying three times, then timing out. The latency spike is real. Its root cause is invisible to every tool the SRE team uses daily.

The release that rolled three hours earlier changed only one thing: the system prompt template was updated for a seasonal campaign. That change shifted the model's output distribution enough that, after six weeks of stability, SKU-code generation began producing fabricated strings in roughly one in forty requests. At 1,800 requests per minute, that is 45 failures per minute, compounding in retry queues until the latency collapsed.

Nothing in the DevOps observability stack detected the shift in output distribution. The canary ran for 20 minutes without an elevated error rate because it evaluated HTTP status codes rather than the semantic correctness of LLM completions. This chapter explains exactly why that gap exists and what it takes to close it.

3.2 Why It Matters for Production Systems

Traditional software is deterministic. A microservice computing shipping cost from weight and destination returns the same result every time those inputs are identical. The entire design of the APM tooling rests on the assumption that "wrong" and "broken" are synonyms.

Machine learning models violate that assumption at a fundamental level. A model executing without error may be producing outputs that are systematically, consequently wrong. Its predictions may have drifted 20 percentage points from validation accuracy. Its confidence scores may be miscalibrated. An LLM may hallucinate structured identifiers at a rate too low to spike error metrics but high enough to degrade user experience significantly. None of these failure modes produces HTTP 5xx codes, and none triggers a threshold alert in Datadog, Dynatrace, or New Relic. They are invisible to infrastructure observability tools by design, not by oversight.

The practical consequence is a false sense of coverage more dangerous than acknowledged ignorance. Engineering teams with mature APM practices underestimate the AI-specific monitoring gap. Specialized tooling is not purchased; the model ships into the same stack as the API it runs behind, and when a model failure eventually surfaces, the team discovers it has weeks of accumulated drift that it cannot reconstruct because inference inputs and outputs were never captured in a queryable store.

Production AI systems fail in four patterns that traditional monitoring cannot detect. Input distribution drift occurs when the statistical properties of data arriving at the model shift from those of the training distribution. Output distribution drift occurs when model predictions cluster differently from those at deployment time, regardless of input changes. Ground-truth performance degradation becomes visible only when outcome labels arrive, revealing declining accuracy against actuals. Semantic drift, specific to generative systems, produces fluent output that is factually wrong or policy-violating in ways no syntactic or infrastructure check can detect.

The team that understands this builds two distinct monitoring coverage planes. The infrastructure plane, owned by SRE, covers uptime, latency, error rates, and resource utilization. The model quality plane, owned by the ML platform team, covers statistical and semantic properties of model behavior using purpose-built AI observability tooling. Both planes need named owners, configured alerts, and tested runbooks. A model quality degradation event will eventually produce infrastructure-layer symptoms; diagnosing those symptoms without model quality context means starting the investigation with the wrong hypothesis.

3.3 DevOps Observability Foundations: Signals, Pillars, and What They Actually Measure

Before understanding what AI observability adds, you need a precise accounting of what traditional DevOps observability covers at the signal level. The three-pillar model of logs, metrics, and traces is genuinely powerful for the problem it was designed to solve. A decade of work in OpenTelemetry standardization, distributed tracing, and SLO management has produced a mature capability for deterministic software systems, and AI observability builds on top of it.

Three Traditional DevOps Observability Pillars

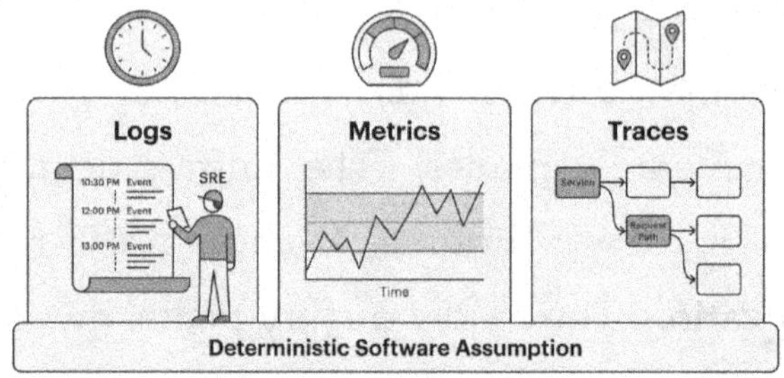

Traditional observability pillars answer whether the system is running — not whether it is right.

Three-pillar telemetry architecture: logs, metrics, and traces in a distributed system instrumented with OpenTelemetry, with signal collection points across the serving stack

What Logs, Metrics, and Traces Cover and Where They Stop

Logs are timestamped records of discrete system events: a request received, a function called, an exception thrown with its stack trace. Their value is forensic. When logs carry trace context, each entry links to the trace ID and span ID of the generating request. For AI systems, logs capture the infrastructure narrative of inference request handling, but nothing about the semantic content of that inference. The log entry that records a successful inference call includes the request ID, latency, and HTTP status code. It does not contain input feature values, output predictions, confidence scores, or any signal of output correctness.

Metrics are time-series measurements of system state: request rate, error rate, latency percentiles, CPU utilization, and queue depth. They power operational dashboards and the SLO calculations that underpin error-budget accounting. That framework applies directly to AI inference services at the infrastructure layer. What metrics cannot measure is the quality of the inference itself. A model can serve 10,000 requests per minute with sub-100-millisecond p99 latency and a zero-error rate, while being wrong on 40 percent of those requests. Every metric is green. The SLO is met. The model is failing.

Traces record a request's path through a distributed system as a tree of spans. OpenTelemetry's GenAI semantic conventions, under active CNCF development, define standard span attributes for AI operations: model name, prompt token count, completion token count, temperature, and finish reason. A trace instrumented with these conventions carries metadata for token-cost attribution and latency breakdown. What the trace does not carry, even with GenAI semantic conventions, is any signal about output quality. The span records that the LLM call was completed, how long it took, and how many tokens it consumed. It does not evaluate whether the completion was factually accurate, policy-compliant, or consistent with historical output distribution.

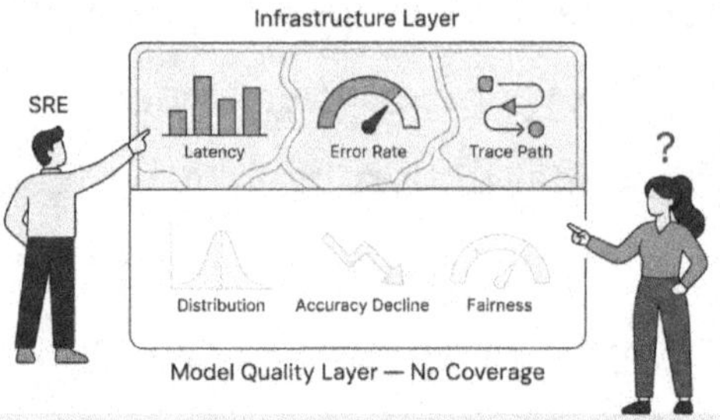

Signal coverage boundary: logs, metrics, and traces covering the infrastructure layer; model quality telemetry covering statistical and semantic properties of model outputs; the gap is the AI observability gap

The Structural Limit: Determinism as the Hidden Assumption

The three pillars share a foundational architectural assumption: correctness is fully determined by whether the system executed without error. For probabilistic models, that assumption does not hold. A model that executes without error and returns a well-formed response has satisfied every criterion the three pillars can evaluate. Whether the response is correct, calibrated, and aligned with the model's intended function requires evaluating the output's semantic and statistical properties, a task entirely outside the scope of infrastructure observability. A model serving predictions through a REST endpoint can have zero errors, sub-100-millisecond latency, and perfect infrastructure health metrics, while its predictions have drifted 18 percentage points from validation accuracy. Every log entry is clean. Every trace completes successfully. The only evidence that the model is underperforming lies in the statistical relationship between its output distribution today and its distribution at training time.

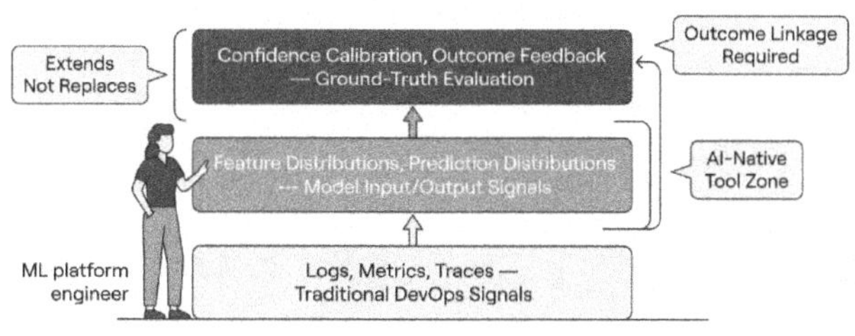

AI observability extends the three pillars with model-quality signal types.

OpenTelemetry distributed trace through an LLM application stack with GenAI semantic convention attributes: model version, token counts, finish reason, and latency spans from gateway through retrieval to completion

3.4 What AI Observability Adds: Signal Types, Instrumentation Patterns, and the Fourth Pillar

AI observability extends the traditional three-pillar stack by adding signal types designed for probabilistic models. Model quality degradation is a statistical phenomenon, not a discrete event. The fourth signal class, model quality telemetry, covers input feature distributions, output value distributions, confidence calibration metrics, evaluator-as-judge scores, and ground-truth feedback loops. Instrumenting for this class requires architectural choices and platform investments that are additive to the existing APM instrumentation.

Deterministic vs Probabilistic System Monitoring

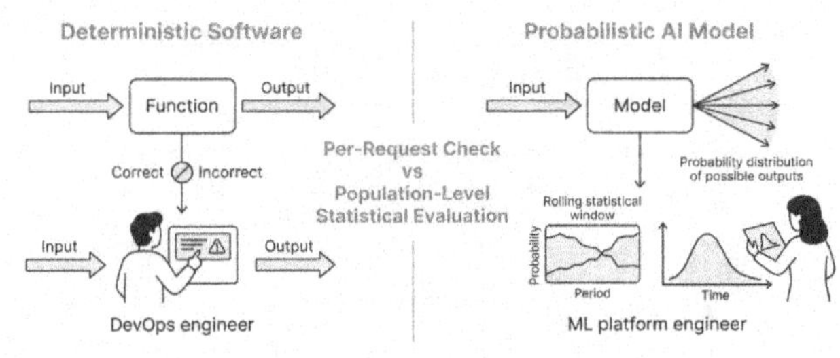

AI correctness is a statistical property measured across many predictions, not individual ones.

Fourth signal pillar: model quality telemetry alongside logs, metrics, and traces, showing data flows from inference endpoints through drift detectors and evaluator pipelines into the unified observability backend

Input and Output Distribution Monitoring

The most reliable early warning signal for model degradation is a shift in the statistical distribution of input features arriving at inference time. Instrumenting for feature drift requires capturing each input feature value per inference request and storing those values for statistical computation over rolling time windows. The standard implementation writes each request's feature vector to a low-latency message queue, with a consumer aggregating feature statistics at configurable window sizes: typically 1 hour, 24 hours, and 7 days. A drift computation job compares the current-window distribution against a reference window captured at deployment time.

For continuous numerical features, the Population Stability Index (PSI) and Kolmogorov-Smirnov (KS) statistic are standard. PSI below 0.1 indicates minimal drift; 0.1 to 0.2 indicates moderate drift; above 0.2 indicates significant drift warranting model review or retraining. For categorical features, chi-squared tests and Jensen-Shannon divergence serve the analogous function. These statistics feed into the central metrics backend, which is queryable alongside infrastructure metrics, with alerting rules routing incidents to the ML platform on-call via PagerDuty, providing context: feature name, drift magnitude, and the reference window being compared against.

Output distribution monitoring tracks shifts in the model's output, regardless of whether input drift is responsible for them. A binary fraud classifier that produced a 3.2 percent positive rate at deployment and now produces a 1.1 percent positive rate warrants investigation, even if no individual input feature PSI threshold has been crossed. For language models, output distribution monitoring operates at a higher level of abstraction: token-level entropy, response-length distributions, refusal rates, and evaluator-as-judge score distributions.

Emergent AI Failure Detection Architecture

Infrastructure metrics are green when LLMs produce confidently wrong outputs.

Output distribution monitoring pipeline: inference outputs captured to the feature store, PSI and KS statistics computed over rolling windows, alert routing into incident management with reference window comparison

LLM-Specific Signals: Evaluator-as-Judge and Prompt Telemetry

Language model observability extends beyond statistical drift detection because LLM outputs are not scalar values or class labels. Quality dimensions that matter in production include factual accuracy, policy compliance, semantic relevance, coherence, and groundedness in retrieved context. None of these can be measured with PSI or KS statistics. The evaluator-as-judge pattern is the standard architecture: a second language model is called on a sampled subset of production outputs to score them against configured quality dimensions, and it receives the original user input, the system prompt, the retrieved context for RAG systems, and the model's response. Scores are emitted as metrics into the central metrics backend and compared against thresholds to generate alerts. The standard pattern is asynchronous evaluation on 1 to 5 percent of production traffic, with the option to increase sampling automatically when drift metrics indicate elevated risk.

Prompt telemetry makes evaluator pipelines operational. Every LLM call should emit a span that includes the prompt template version identifier, the rendered prompt or a privacy-safe hash, retrieved context identifiers for RAG systems, the completion token count, model version, and a unique evaluation ID linking the production call to any downstream evaluator call. This span structure enables engineers to correlate quality degradation with specific prompt template versions, retrieval pipeline behaviors, or model version changes, giving the on-call engineer a structured investigation path. The AI monitoring platform integrates with incident management workflows by firing alert webhooks to PagerDuty or ServiceNow when a drift threshold is crossed, or an evaluator score falls below its minimum.

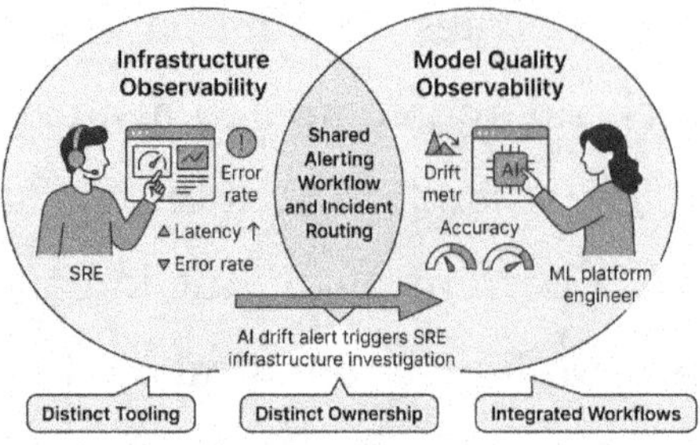

Infrastructure SREs and ML platform teams each own a distinct observability coverage area.

LLM observability stack: prompt telemetry spans linked to async evaluator-as-judge pipeline, with quality scores flowing into the metrics backend and alert routing with pre-populated incident context

3.5 The Observability Gap in Practice: What to Instrument, How to Sequence, and the Maturity Spectrum

Building monitoring coverage requires sequencing instrumentation work and advancing from a reactive baseline toward a predictive posture.

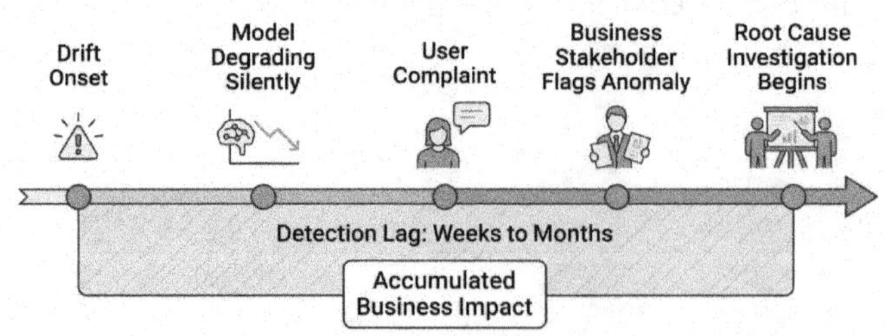

Reactive detection allows weeks of silent degradation before discovery.

AI observability maturity spectrum from reactive (user-reported failures, no inference capture) through proactive (automated drift alerts, reference windows configured) to predictive (leading indicators, upstream pipeline monitoring)

Inference Capture and Reference Distributions: The Foundation Layer

The single most valuable first step is inference capture: instrumenting the model serving layer to write each inference request's inputs and outputs to a queryable store at serving time. Without this capability, no downstream drift detection, evaluator pipeline, or quality monitoring is possible. The architecture is a middleware layer that intercepts each request and response, extracts relevant fields, and asynchronously writes them to a message queue, which a consumer then writes to a persistent store: a feature store, a data warehouse table, or a purpose-built prediction log. The write must be non-blocking, with acceptable additional latency under 5 milliseconds at p99.

Once inference capture is operational, the next priority is to establish a reference distribution: the statistical fingerprint of the model's inputs and outputs, computed from a window of stable production traffic shortly after deployment. Storing the reference as a versioned artifact in the model registry ensures the drift detection configuration is reproducible and auditable. With inference capture and a reference distribution in place, the team implements the first drift-detection job: a scheduled hourly computation that computes PSI and KS statistics, compares them against the reference distribution, and emits the results as Prometheus metrics.

Canary and Shadow Deployments: Extending Release Engineering for Model Quality

Canary and shadow deployment patterns apply directly to model releases but must be extended with model-quality evaluation logic. For AI systems, canary evaluation must check output-distribution divergence between the canary and baseline, compare evaluator-as-judge scores for LLM systems, and compare ground-truth performance if labels are available. A canary that passes infrastructure health checks but shows significant divergence in output distribution should not be promoted without investigation. Shadow deployments route a copy of production traffic to the new model version without serving the shadow outputs to users, allowing thorough quality comparison before any production traffic is committed. The observability instrumentation for both patterns must use the model version identifier as a dimension on all metrics, logs, and traces. OpenTelemetry resource attributes are the correct mechanism: setting model version as a resource attribute on the tracer and meter causes it to propagate automatically to all spans and metrics emitted during that deployment's request handling.

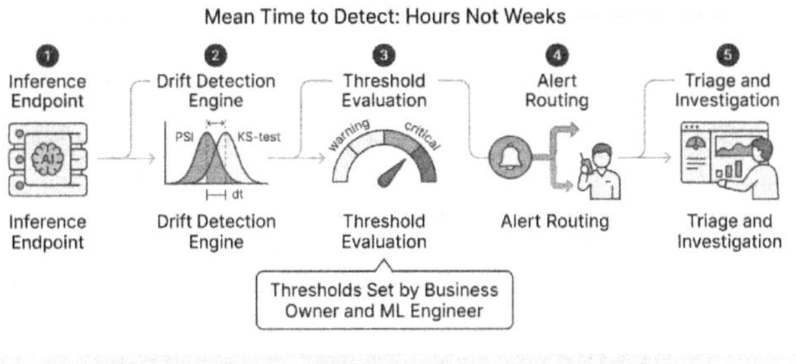

Proactive monitoring reduces detection lag from weeks to hours.

Canary model release with dual evaluation: infrastructure health checks (latency, error rate) plus model quality checks (output distribution divergence, evaluator scores), gating the traffic promotion decision

Reactive, Proactive, Predictive: The Maturity Progression

The realistic progression follows a maturity spectrum with three distinct operational postures. At the reactive posture, model failures are discovered through user reports or business KPI anomalies. MTTD is typically measured in weeks or months, and remediation begins with root-cause analysis from first principles. This is the default state of any AI deployment that did not invest in monitoring before go-live. The proactive posture requires three capabilities: operational inference capture, automated drift detection with configured thresholds, and named on-call ownership of model quality alerts. Drift alerts fire based on statistical thresholds rather than user complaints, and MTTD drops from weeks to hours. For any model managing consequential decisions, the proactive posture is the minimum viable monitoring coverage.

The predictive posture adds leading-indicator monitoring: tracking upstream data-pipeline health signals, feature-variation trends, and model-specific precursor patterns that historically precede accuracy degradation before drift statistics cross alert thresholds. Building predictive observability requires at least six to twelve months of continuous monitoring data and documented drift events with their upstream precursors. A demand forecasting model may show leading indicators in pipeline freshness metrics 48 hours before output drift crosses the alert threshold. An NLP classifier may show leading indicators in shifts in the input token-length distribution before the output class distribution drifts. A named observability lead, owning the monitoring platform, evaluation frameworks, alert configuration, and incident review across all production models, becomes a full-time position at this maturity level.

AI Observability Maturity Progression

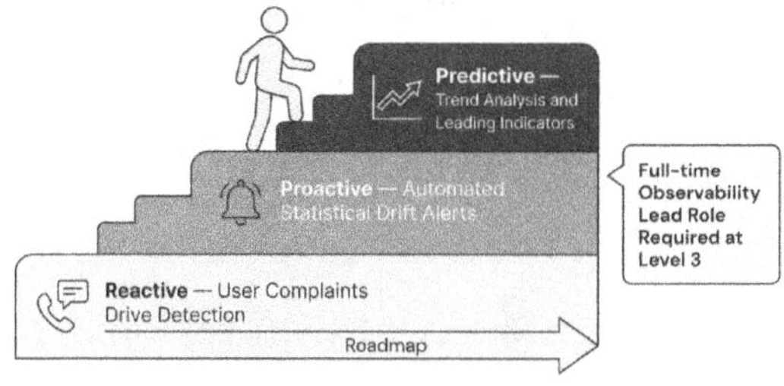

Every enterprise starts reactive; the target state is predictive intelligence.

Predictive observability: upstream data pipeline signals and feature variance trends feeding a pre-alert detection layer before drift statistics cross threshold, with historical drift event correlation from the monitoring archive

3.6 Technical Team Checklist: DevOps-to-AI Observability Transition

- **Action:** Audit every production AI system to confirm inference inputs and outputs are captured to a queryable store; instrument the serving middleware for any system without capture. **Owner:** ML Platform Eng. **Done when:** 100 percent of production model endpoints write inference records, verified by a daily data-freshness check with a sub-1-hour staleness SLO.
- **Action:** Compute a reference distribution for each production model from a 7-day window of stable post-deployment traffic and store it as a versioned artifact in the model registry. **Owner:** ML Engineer. **Done when:**

Each model in the registry has a linked reference distribution artifact with a version that matches the deployed model, accessible via the registry API.

- **Action:** Implement automated PSI and KS-statistic computation for continuous features and Jensen-Shannon divergence for categorical features on a 1-hour schedule, emitting results as Prometheus metrics queryable in Grafana. **Owner:** ML Platform Eng. **Done when:** Drift metrics for all monitored features carry at least 30 days of history, and alerting rules are configured with per-feature thresholds documented in the model's observability runbook.
- **Action:** Configure PagerDuty routing to create high-urgency incidents from model quality alert webhooks with ML platform on-call as primary responder, each alert pre-populated with model name, feature name, drift magnitude, and a link to the relevant Grafana panel. **Owner:** SRE / On-Call. **Done when:** A synthetic drift threshold breach in staging generates a PagerDuty incident within 2 minutes with correct context fields.
- **Action:** Write and peer-review on-call runbooks for each class of model quality alert, covering investigation steps, root cause hypotheses in priority order, remediation actions, and escalation path. **Owner:** ML Engineer. **Done when:** Each runbook has been reviewed by an ML platform engineer who was not the author, and a

tabletop exercise has been completed using a synthetic incident scenario.

- **Action:** Instrument all LLM serving endpoints with OpenTelemetry spans carrying GenAI semantic convention attributes and configure an async evaluator-as-judge pipeline sampling at least 2 percent of production completions for factual accuracy, grounding, and policy compliance. **Owner:** ML Platform Eng. **Done when:** Evaluator scores are queryable at 1-hour granularity, and an alerting rule fires when the 24-hour rolling average falls below configured thresholds.
- **Action:** Extend canary deployment promotion criteria for all model releases to include output distribution divergence below PSI 0.1, evaluator score delta within configured bounds, and no open model quality incidents from the canary version. **Owner:** ML Platform Eng. **Done when:** The CI/CD pipeline blocks promotion if any quality-gate criterion fails, with the gate result logged in the deployment audit trail.
- **Action:** Verify OpenTelemetry GenAI semantic convention compliance for every new AI system before production go-live, covering span attribute completeness, trace context propagation, and metric naming adherence. **Owner:** Architect. **Done when:** The production readiness review includes a GenAI semantic convention

compliance check as a required pass criterion verified by an automated test in CI.

3.7 Team Conversation Starters

- When we say a model is "in production and monitored," what specifically are we measuring? Are we covering the serving infrastructure, the model's output distribution, or both, and do we have tested runbooks for each alert type?
- If a key input feature's PSI crossed 0.2 right now, who gets paged, what does the runbook tell them to do first, and what is our target MTTD for a feature drift event?
- For each LLM endpoint we operate, can we answer "what is the current evaluator-as-judge factual accuracy score" without manual work? If not, what specific gap is blocking us?
- What is our actual MTTD for a model quality failure based on the most recent real incident, and what single capability investment would cut it in half?
- When we promote a new model version through Canary, are we evaluating output distribution divergence and evaluator scores alongside error rate and latency, or only checking infrastructure health metrics?
- Do our on-call runbooks for model quality alerts contain enough context for an engineer who was not involved in

building the model to follow them end to end, and when did we last test that assumption in a tabletop exercise?

3.8 Key Takeaway

Traditional DevOps observability covers the infrastructure half of an AI production system with precision and maturity. Logs, metrics, and traces answer the question "Is it running?" completely. They answer the question "Is it right?" not at all. Production AI systems require a fourth signal class, model quality telemetry, covering input and output distribution statistics, confidence calibration, evaluator-as-judge scores for generative systems, and ground-truth feedback loops. These signals must be captured at inference time, computed against deployment-time reference distributions, and wired into the same incident management workflow as infrastructure alerts, with named on-call ownership and tested runbooks. The gap between APM coverage and AI coverage is not a vendor limitation; it is an architectural distinction that requires deliberate instrumentation work to close.

4 Drift, Decay, and Degradation: What Goes Wrong in Production AI

4.1 Opening Scenario

It starts as a Slack message from the fraud ops team on a Tuesday morning: "Our catch rate feels low this week." By the time it reaches Priya, the on-call data scientist, the team has escalated it to a Sev-2 ticket. The model is a gradient-boosted tabular classifier scoring card-not-present transactions for fraud risk. It has been running in production for seven months without a retraining cycle and without any monitoring configuration beyond latency and error rate.

Priya pulls up the prediction logs and plots daily recall over the past two months. The slope is consistent: recall has been bleeding from 0.87 down to 0.74 over six weeks. No alert fired because no alert was wired. The feature-level PSI checks that the platform team had planned to configure were never implemented after a sprint reprioritization. She pivots to the feature store snapshots. The transaction amount and merchant category code look clean. But the velocity feature, which counts distinct merchant categories a card has touched in the past 72 hours, has shifted meaningfully. PSI computed retrospectively against the training reference window is 0.31, well above the 0.2 threshold that signals a significant distribution shift.

She traces the velocity feature back to its upstream pipeline and finds the root cause: a data engineering team pushed a schema change to the raw events table six weeks ago. A column rename silently changed how the feature transformation query joined events, causing undercounting of category touches for transactions arriving in certain time-zone offsets. The model has been receiving a corrupted velocity signal for exactly as long as recall has been declining.

The fix takes four hours. Detection, retrospective forensics, and incident write-up take three days. The failure was entirely preventable by adding a PSI check on the velocity feature, tied to an alerting threshold. This chapter covers this class of failure and the four other distinct failure modes that a single PSI check would not have caught.

4.2 Why It Matters for Production Systems

Every production ML model is a frozen artifact operating in a world that keeps moving. At training time, the model encodes statistical relationships between input features and target outcomes as they existed in historical data. Once deployed, those learned relationships are fixed while the world continues to evolve: user behavior shifts, upstream data sources change, adversarial actors adapt, and internal systems are modified. The widening gap between what the model learned and what the world looks like now is the structural source of every failure mode covered in this chapter.

The engineering-correct view treats degradation as a taxonomy of distinct failure modes, each with its own causal mechanism, detection method, and remediation path. Collapsing them into one alert bucket is like running a distributed microservices cluster with a single "system unhealthy" alert: technically possible but practically unactionable. Covariate shift occurs when the marginal distribution of one or more input features diverges from the training distribution, placing inputs in a region of feature space sparsely sampled at training time. Label drift occurs when the base rate of the target class changes; if a fraud model's actual fraud rate shifts from 0.8 to 0.3 or 2.1 percent, calibrated probability outputs mislead downstream decision logic. Concept drift is a change in the functional relationship between inputs and the target, with no detectable change in input distributions. Feature pipeline corruption arises from changes in the systems that produce the features the model consumes; a schema migration, a vendor API change, or a time zone-handling bug can alter the statistical properties of engineered features without causing a pipeline error.

For embedding-based models, embedding drift adds a further dimension: a sentence encoder fine-tuned on 2022 data will produce different embedding geometries for queries containing product terminology that emerged in

2024, and vector similarity metrics against a static index will degrade silently. The observability platform must instrument both the raw input boundary and the embedding output boundary to distinguish input-vocabulary drift from encoder-model drift, because they require different remediation. Teams that understand pipeline corruption implement data contracts between the feature engineering team and the model serving layer with automated schema and distribution validation at the boundary.

4.3 Measuring Input Distribution Shift: PSI, KS-Statistic, KL Divergence, JS Divergence, and MMD

The foundational engineering problem in drift detection is operationalizing "has the distribution of this feature changed?" as a computable, thresholdable signal wired into an alerting pipeline. Several statistical measures address this question, each with different computational properties and practical trade-offs.

Population Stability Index and Reference Window Design

The Population Stability Index is the most widely deployed drift metric in production ML platforms. PSI quantifies divergence between a reference distribution and a current distribution by binning both and aggregating: PSI equals the sum over K bins of (P_current minus P_reference) times the natural log of (P_current divided by P_reference). Conventional thresholds: PSI below 0.1 indicates no meaningful drift; 0.1 to 0.2 indicates moderate drift; above 0.2 requires investigation. Production implementations should use equal-frequency binning derived from the training reference distribution and fix those bin boundaries at deployment time so that all future computations are comparable. A static reference window that uses the training set directly measures drift from the model's learned distribution, but also fires alerts for expected seasonal patterns that the model handles correctly. A sliding reference window of 30 to 90 days adapts to slow drift but fails to detect gradual drift accumulation. A golden dataset reference uses a manually validated sample representing the ideal operating distribution; deviations above a threshold trigger a quality gate that blocks model output from automated decisions, making it appropriate for models under regulatory oversight.

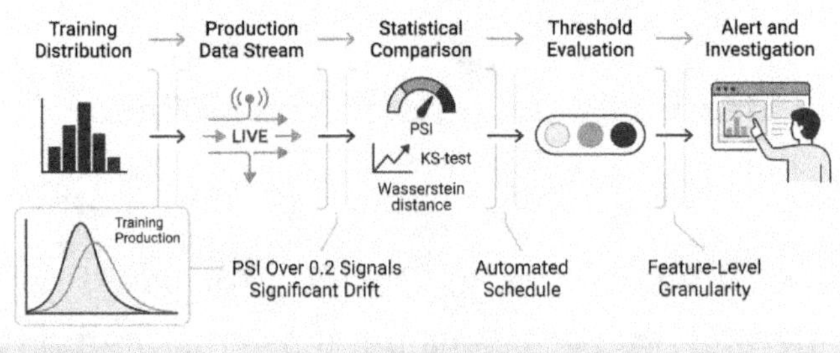

Drift detection pipelines compare production distributions against the training baseline continuously.

PSI computed per feature against training reference distribution; bin-level divergence heatmap highlighting features that exceed the 0.2 threshold

KS-Statistic, KL Divergence, JS Divergence, and Wasserstein Distance

The Kolmogorov-Smirnov test operates on empirical cumulative distribution functions. The KS statistic is the maximum absolute difference between the two CDFs, returning a p-value indicating whether the two samples are likely drawn from the same distribution. The KS test is more sensitive to local shifts in the tails than PSI and is most appropriate for features where tail behavior is predictively significant. KL divergence measures the information lost when the reference distribution approximates the current distribution: the sum over x of P(x) times the log of P(x) divided by Q(x). KL divergence is asymmetric and sensitive to bins with near-zero probability in Q, so it requires smoothing. Jensen-Shannon divergence addresses asymmetry by computing the average KL divergence between each distribution and its mixture; it is bounded between 0 and 1 using base-2 logarithms and is the preferred choice for heavy-tailed or sparse distributions. Wasserstein distance measures the minimum probability-mass transport required to transform the reference distribution into the current distribution; for univariate distributions, the

Wasserstein-1 distance is the L1 distance between their quantile functions. Wasserstein distance is appropriate for features such as embedding norms or latency distributions, where gradual shifts in central tendency matter as much as changes in the tails.

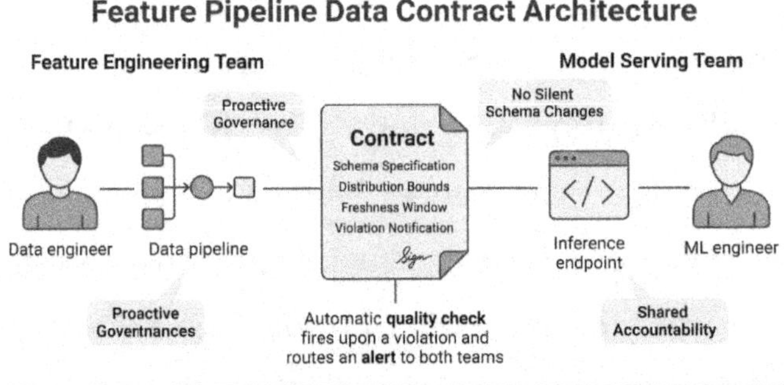

Comparison of KS statistic, JS divergence, and Wasserstein distance curves for the same feature over a 90-day production window; divergent alarm timing illustrates each metric's sensitivity profile

Maximum Mean Discrepancy and Embedding Drift Detection

Maximum Mean Discrepancy extends drift detection into high-dimensional spaces where univariate metrics are insufficient. MMD measures the distance between two distributions in a reproducing kernel Hilbert space using a radial-basis-function kernel, without requiring dimensionality reduction or binning. Libraries, including Alibi Detect and Evidently, provide optimized MMD estimators with bootstrap-based p-value computation. Random Fourier feature approximations reduce the naive quadratic cost to linear, making real-time embedding drift monitoring feasible at production scale. For production embedding stores used in RAG pipelines, embedding drift manifests as degradation in nearest-neighbor retrieval quality: as the query distribution drifts away from the index corpus, mean cosine similarity between query embeddings and top-k retrieved neighbors decreases, and retrieval precision degrades. Instrumenting the retrieval layer to emit similarity-score histograms alongside latency telemetry, this signal is sent into the observability pipeline. The platform must instrument both the raw-input boundary and the embedding-output boundary to distinguish input-vocabulary drift from encoder-model drift. Input vocabulary drift suggests retraining the downstream task model. Encoder model drift requires rebuilding the vector index against the new embedding geometry before any retraining discussion is warranted.

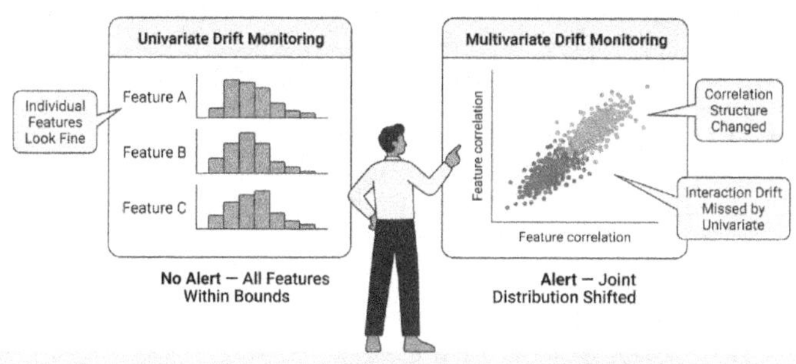

Univariate monitoring misses drift in feature interaction spaces critical to model performance.

MMD scores over rolling 7-day windows for a sentence encoder embedding space; upward trend preceding retrieval precision drop illustrates embedding drift as a leading indicator

4.4 Model Decay and Performance Degradation: Detection Patterns for Tabular and Generative AI

Input distribution monitoring catches drift at the model's front door. The metric that ultimately matters is whether predictions remain accurate enough to justify continued deployment. Ground truth arrives late, arrives incompletely, or, for generative AI, requires evaluation machinery that is itself a complex system component.

Tracking Model Decay with Ground Truth Labels and Sliding Evaluation Windows

For classification models with ground-truth labels available with known lag, the core degradation signal is a rolling evaluation of predictions against confirmed outcomes. The implementation requires a label-joining pipeline that matches production predictions to the eventual ground-truth labels and feeds the joined records into a metrics computation layer. The metrics depend on the model's operational context: recall and precision for imbalanced classification, calibration error for probability-threshold routing, and RMSE or mean absolute error for regression models. Performance degradation thresholds should be documented in the model card as explicit SLOs. An SLO of the form "recall must remain above 0.82 over any rolling 14-day window" is queryable in the same dashboard that displays latency and error rate. When breached, the on-call runbook specifies the escalation path, initial diagnostic queries, and criteria for triggering emergency retraining versus rollback. Output distribution PSI, computed on the model's score histogram rather than on input features, is a useful leading indicator of concept drift in long-lag settings where ground truth may not arrive for 12 to 18 months.

Concept Drift Detection Framework

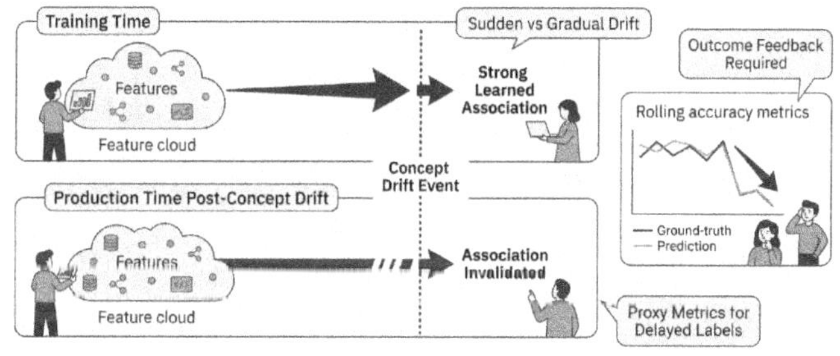

Concept drift is invisible to feature distribution tests — only outcome feedback reveals it.

Recall SLO tracking over rolling 14-day windows; dashed threshold line and breach annotation show the point at which on-call escalation triggers.

Concept Drift: Covariate Shift versus Label Shift versus Posterior Shift

Covariate shift is a change in P(X) without a change in P(Y|X). Under pure covariate shift, a calibrated model remains correct in expectation; performance degrades only because regions of input space that were heavily populated during training were underrepresented in the training data. PSI and KS tests on input features catch it directly. Label shift is a change in P(Y) without a change in P(X|Y). A fraud model calibrated for a 0.8 percent fraud rate will be miscalibrated if the actual fraud rate shifts to 0.4 or 1.5 percent, even if fraudulent transactions look the same as they did at training time. The Black-Box Shift Estimation framework provides a principled method for detecting label shift using only model predictions, making it useful in long-lag settings. Posterior shift is a change in P(Y|X): the same input values now correspond to a different distribution over outcomes. This is true concept drift and directly invalidates the model's learned function. Sudden concept drift occurs when a discontinuous event abruptly changes the relationship between features and outcomes. Gradual concept drift requires trend analysis across extended rolling windows and monitoring the output confidence distribution as a proxy for posterior stability.

Model Decay Lifecycle and Retraining Trigger

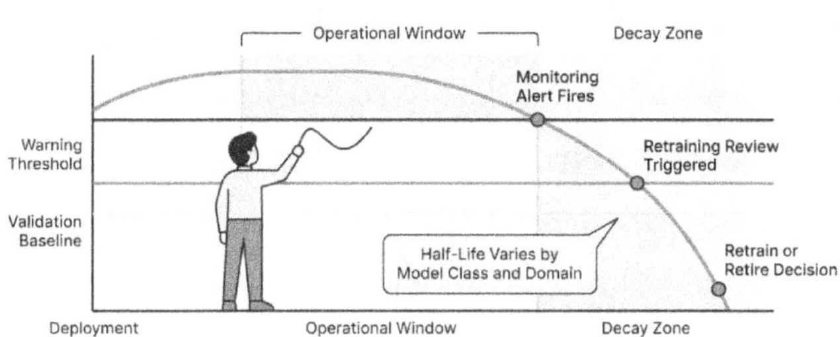

Every production model has a characteristic decay half-life that monitoring must track.

Three-panel diagram distinguishing covariate shift, label shift, and posterior shift; detection method and remediation path annotated for each type

LLM Hallucination Drift, Prompt Sensitivity, and Context Window Degradation

Large language models present degradation patterns that have no direct analog in tabular classification. Hallucination rate is the fraction of model outputs containing factually incorrect, plausible-sounding claims. As user query patterns evolve toward topics underrepresented in fine-tuning data, the hallucination rate increases without any change in model weights. Tracking hallucination rates requires an evaluator-as-judge pipeline: a faster LLM or fact-checking model that scores each production output for factual consistency against a verified reference. Tools such as Langfuse, LangSmith, and Ragas provide frameworks for this pipeline. The rolling average over a 1,000-query window is stable enough to support threshold-based alerting. Prompt sensitivity drift occurs when changes in prompt construction logic, or gradual shifts in how users phrase queries, cause the model to produce systematically different output distributions for semantically equivalent requests. Detection requires a canonical test suite: a fixed set of 50 to 200 representative queries evaluated against the live model on a scheduled cadence, with regression treated as a deployment-blocking signal.

Context window degradation is a RAG-specific failure mode. As the document corpus grows, ages, or is updated, retrieved context quality degrades in ways that cascade into degraded LLM output. The primary monitoring signals are the mean cosine similarity between the query and retrieved chunks, context relevance scores from the evaluator pipeline, and faithfulness scores measuring whether the model output is grounded in retrieved context rather than in model weights. Co-locate these signals with LLM output quality metrics on the observability platform so the on-call engineer can see the complete picture in a single view.

LLM Failure Mode Monitoring Coverage

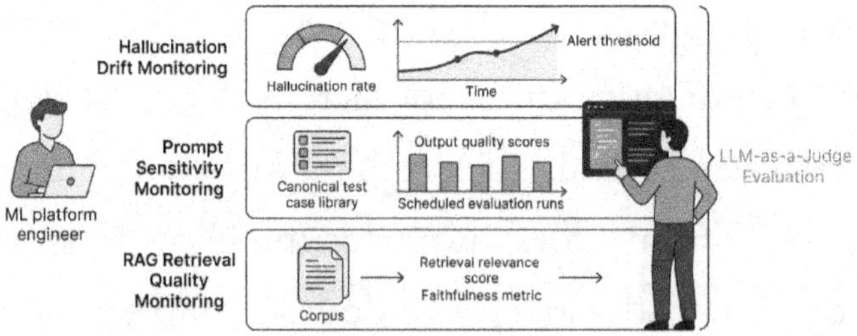

LLM monitoring requires three distinct coverage tracks not present in traditional ML observability.

LLM eval harness architecture showing canonical test suite, evaluator-as-judge pipeline, and metric emission path into the centralized telemetry backend

4.5 Detection Patterns, Reference Window Design, and Incident Response Integration

Detecting drift is a solved problem in the sense that the statistical tools exist and are well-understood. The harder engineering problem is designing the detection system so that drift signals reach the right people with the right context at the right time. A drift alert firing into a generic monitoring channel with no diagnostic context is barely better than no alert.

Alert Routing, Runbook Integration, and Escalation Design

The metric computation layer runs PSI, KS tests, output distribution checks, and performance metric evaluations on a scheduled cadence. A real-time fraud model warrants 15-minute PSI checks on the highest-signal features; a monthly demand forecasting model can tolerate daily checks. All metrics should be emitted as time-series data into the central metrics backend, so drift signals are queryable alongside latency and throughput. Threshold configuration should be stored in a versioned configuration file so changes are tracked in version control and are auditable. Alert routing should leverage the same incident management infrastructure used for production service alerts: a PSI breach on a high-stakes model should create a PagerDuty or Opsgenie incident with the feature name, PSI value, reference window dates, a link to the distribution timeline, and the suggested initial diagnostic query.

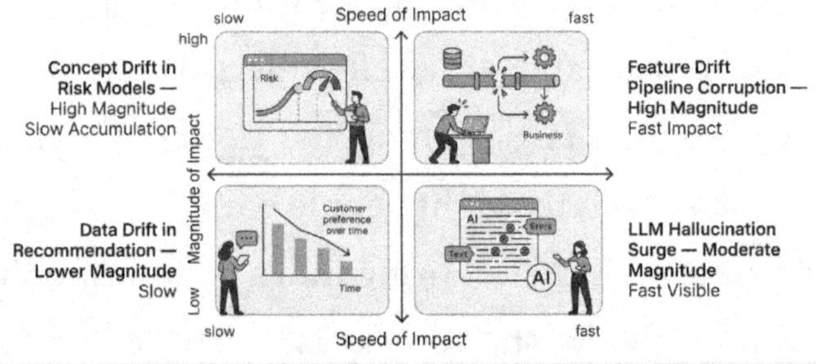

Failure mode impact profiles determine monitoring priority and alert threshold tightness.

Drift alert routing pipeline from metric computation through threshold evaluation to incident creation; diagnostic context payload fields annotated at each stage

Reference Window Design and Remediation Patterns

A static reference window using the training set directly measures drift from the model's learned distribution. A sliding reference window of 30 to 90 days adapts to gradual drift in periodically retrained models but fails to detect the accumulation of drift. A golden dataset reference uses a manually curated, validated sample that represents the ideal operating distribution; deviations trigger a quality gate that blocks model output from being used in automated decisions. Feature store snapshots should be versioned and retained alongside model registry entries to enable retrospective drift analysis.

Retraining triggers should be specified as observable metric thresholds in the model's on-call runbook: "if rolling 14-day recall drops below 0.82 or if PSI on any tier-1 feature exceeds 0.25, open a retraining review within 4 hours." The retraining review should evaluate whether degradation is attributable to input drift, pipeline corruption, or concept drift, because each requires a different remediation. Rolling back to a previous model version is the fastest remediation when a stable version is available in the model registry. Kill switches are required for any model making autonomous decisions: a feature flag routing traffic to a fallback rule system or human review queue when a drift threshold is exceeded, operable without a deployment.

Drift Threshold Calibration Process

Threshold calibration is a three-party business decision, not a statistical choice alone.

Three reference window strategies side by side: static, sliding, and golden dataset; alert behavior under gradual drift illustrated for each strategy

Failure Mode Detection Gap Analysis

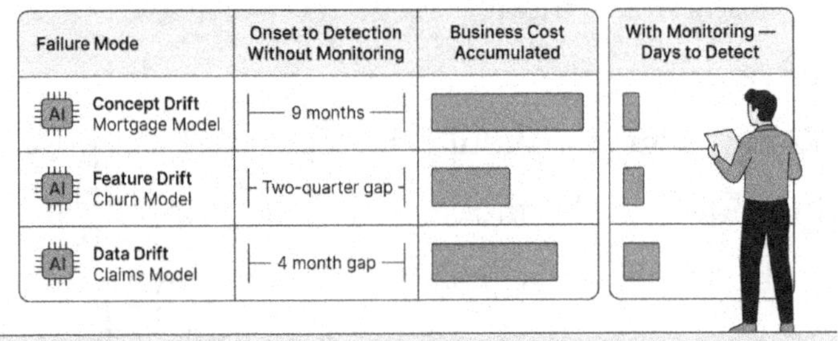

Failure Mode	Onset to Detection Without Monitoring	Business Cost Accumulated	With Monitoring — Days to Detect
AI Concept Drift Mortgage Model	9 months		
AI Feature Drift Churn Model	Two-quarter gap		
AI Data Drift Claims Model	4 month gap		

Every failure mode in this analysis was detectable at onset — monitoring was the missing element.

Remediation decision tree from drift alert through retraining review, rollback, and kill switch activation; ownership and escalation path annotated at each decision node

4.6 Technical Team Checklist: Drift Detection and Response

- **Action:** Instrument PSI computation for every tier-1 input feature against a versioned reference window, emitting results as time-series metrics into the central observability backend at a cadence appropriate to model sensitivity. **Owner:** ML Platform Eng. **Done when:** PSI values for all tier-1 features are queryable in the same dashboard as model latency and error rate, with a documented reference window strategy in the model card.
- **Action:** Implement KS-statistic monitoring for continuous features where tail behavior is predictively

significant, with p-value thresholds calibrated to the acceptable false-positive rate for the model risk tier. **Owner:** ML Engineer. **Done when:** KS tests run on every model with continuous features, and results appear in the drift dashboard alongside PSI metrics.

- **Action:** Deploy MMD-based multivariate drift detection for any model whose predictive power depends on feature interaction structure, including gradient-boosted ensembles, deep tabular models, and embedding-based retrievers. **Owner:** ML Platform Eng. **Done when:** MMD score time series is queryable per model, and an alert fires when MMD exceeds the calibrated threshold.
- **Action:** Define output distribution PSI monitoring for every production model as a concept drift leading indicator, tracking the model's score histogram against a reference window. **Owner:** ML Engineer. **Done when:** Output distribution PSI is tracked against a documented threshold, and a breach triggers a Sev-2 incident with a runbook link in the alert payload.
- **Action:** Establish a data contract between each feature engineering pipeline and its downstream model serving layer, specifying expected schema, distribution bounds, freshness SLA, and automated quality gate behavior on contract violation. **Owner:** Data Eng. **Done when:** Contract violations trigger an automated alert before the

model consumes affected features, tested end-to-end in staging.

- **Action:** Build a canonical eval harness for every LLM deployment with 50 to 200 representative test cases running on a cadence of no more than 6 hours, emitting pass rate and evaluator-as-judge quality scores into the central metrics backend. **Owner:** ML Engineer. **Done when:** Canonical test suite regression is a deployment-blocking signal in CI/CD, and a rolling average below the quality threshold creates a paging incident.
- **Action:** Document per-model retraining triggers as observable metric thresholds in each model's on-call runbook, specifying the feature PSI threshold, the performance SLO breach threshold, and the escalation path, including rollback authorization. **Owner:** ML Platform Eng. **Done when:** The on-call runbook for each production model includes a retraining trigger section, signed off by both the ML Engineer and the SRE during the pre-deployment review.
- **Action:** Implement versioned feature store snapshots retained alongside model registry entries so that historical PSI retrospective analysis is possible for any production model going back to its deployment date. **Owner:** Data Eng. **Done when:** Any engineer can query the PSI value for any feature at any historical date using a documented dashboard query.

- **Action:** Configure kill switch feature flags for all models, making autonomous decisions operable by the on-call engineer without a deployment, routing traffic to a documented fallback system when triggered. **Owner:** SRE / On-Call. **Done when:** Kill switch is tested in staging, fallback routing is verified end-to-end, and the trigger procedure is documented in the incident runbook.
- **Action:** Maintain a threshold change log as a versioned governance artifact for each production model, recording each PSI, KS, or SLO threshold change with the business rationale and the approving engineer. **Owner:** Governance Eng. **Done when:** Threshold change log is linked from the model card and included in the quarterly model health review audit package.

4.7 Team Conversation Starters

- If the velocity feature on our fraud model were corrupted by a schema change today, which specific PSI check would fire first, and how long would detection take under our current monitoring configuration?
- Have we distinguished between covariate shift monitoring, label shift monitoring, and concept drift monitoring in our model health dashboards, or are we treating all drift as the same signal with the same response path?

- What is the reference window strategy for each production model, and has that choice been deliberately documented in the model card rather than defaulted to whatever the monitoring tool uses?
- For our LLM deployments, what is the current mean time to detect a hallucination rate increase of 5 percentage points over baseline, and what would get that below 24 hours?
- Are our drift thresholds calibrated against the cost asymmetry between false positive investigations and missed degradation events, and has that calibration been reviewed in the last 90 days?
- Can every on-call engineer execute a model rollback or kill switch activation without escalation, and has that procedure been tested in staging in the past 30 days?

4.8 Key Takeaway

Drift, decay, and degradation are not interchangeable names for the same problem. Covariate shift, label drift, concept drift, feature pipeline corruption, and embedding drift each has distinct causal mechanisms, detection methods, and remediation paths. A production ML observability platform that treats them as a single model-degradation bucket will systematically fail to detect failure modes it is not explicitly instrumented to catch. The engineering requirement is a layered detection architecture:

input distribution monitoring using PSI, KS-statistic, and MMD at the feature boundary; output distribution monitoring as a concept drift proxy; performance SLO tracking against ground truth where available; and a continuously running eval harness for generative AI components. Wire all of them into the same alerting and incident management infrastructure used for service reliability, and back every alert with a runbook that specifies ownership, diagnostic steps, and remediation authority before the model reaches production.

5 The Five-Layer AI Observability Stack for Production Systems

5.1 Opening Scenario

Priya joined the war-room call at 11:47 PM from her home office, laptop fan already spinning. The incident commander had paged her twenty minutes earlier: the customer-facing chatbot for a large retail bank was producing responses that explicitly named competitor interest rates, a policy violation the compliance team had supposedly resolved three months ago. The channel was full of engineers arguing about prompt templates, firewall rules, and whether a recent model upgrade was to blame.

Priya ran the incident layer by layer, from the bottom of the stack upward. Infrastructure first: GPU utilization was elevated at 91 percent, above the 85 percent warning threshold, and inference queue latency was at p99 of 4.2 seconds, nearly double the SLO baseline. Stress, but stress alone does not cause policy violations.

Data layer: the freshness SLO for the retrieval augmentation index was green, but a schema validation alert from forty minutes earlier had been auto-acknowledged by a bot. She opened it: the document ingestion job had pulled a new batch of competitor rate sheets from a knowledge management system that a content team had accidentally added to the ingestion scope. Schema valid, row counts normal, data quality SLIs all passed. The content was wrong.

Model layer: output distribution stable, confidence scores within range, no drift signal anywhere. The model was doing exactly what it was designed to do: ground responses in retrieved context. The retrieved context was the problem.

Layer 4 confirmed it. The online evaluator had flagged fourteen policy-violating responses in the last hour, but the alert had been routed to a queue no one was watching. The governance layer had the attestation log showing the retrieval index had been updated outside the change-management window.

Priya had the root cause in eleven minutes: one misconfigured data source, one unmonitored alert queue, and a governance gap around index update approvals. Without all five layers instrumented and connected, this incident would have run until morning.

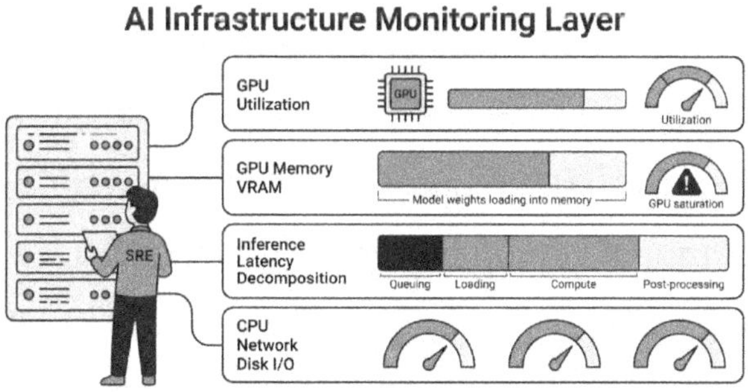

AI Infrastructure Monitoring Layer

AI infrastructure monitoring extends traditional APM with GPU and inference-specific decomposition.

War-room trace: GPU saturation, retrieval index schema alert, and policy-violation eval flag correlated via trace ID

5.2 Why It Matters for Production Systems

A production AI system is a layered system in which infrastructure, data pipelines, model inference, output quality evaluation, and governance controls operate on different timescales, emit different signal types, and fail in different ways. Instrumenting only one or two of those layers yields an observability program with structural blind spots that allow incidents to run for an hour before anyone traces the root cause.

AI system failures rarely announce themselves as clean, categorical errors. A traditional web service returns a 500 status code when it fails. An AI system degrades. Output quality erodes gradually, feature distributions drift over weeks, retrieval indexes silently absorb bad documents, and model confidence scores shift in ways that appear to be noise until they cross a threshold that matters to the business. The failure mode is statistical and semantic, not binary.

Consider the gaps created by partial instrumentation. Infrastructure-only monitoring gives you GPU utilization, memory pressure, and request latency, but nothing about whether the model is making correct predictions. A model can serve thousands of wrong answers at a p50 latency of 200 milliseconds, with every infrastructure metric looking perfect. Add data pipeline monitoring, and you gain freshness SLOs and schema validation, but a perfectly fresh, schema-valid feature set can still represent a population shift that degrades model accuracy. Add model distribution monitoring, and you can detect output drift, but not whether outputs are semantically correct or policy-compliant. For LLMs, correctness is not a statistical property: a model can produce fluent, confident output that is factually wrong or violates a regulatory boundary. Add evaluation without governance, and you have real-time quality signals but no immutable audit trail and no mechanism for proving to a regulator that you detected a violation and acted on it.

The five-layer architecture exists because each layer answers a different diagnostic question. Layer 1: Is the serving environment healthy? Layer 2: Is the input data fresh, valid, and statistically consistent with the training distribution? Layer 3: Has the model's output distribution or calibration changed relative to the validation baseline? Layer 4: Are individual outputs correct, safe, and policy-compliant? Layer 5: Is there a verifiable, immutable record of every model version, deployment decision, eval result, and policy attestation?

Trace correlation IDs connect the layers. OpenTelemetry is the architectural mechanism: when every layer emits telemetry with a shared trace context, the five-layer stack becomes a navigable observability plane rather than a set of five separate dashboards. Enforce one contract from day one: every layer must emit a trace context that the next layer can consume. Without the data layer, you spend the first thirty minutes of every model quality incident ruling out pipeline problems manually. Without the eval layer, policy violations surface from customer complaints. Without the governance layer, answering the regulator's question about which model version was running at a specific time requires a manual audit that takes days.

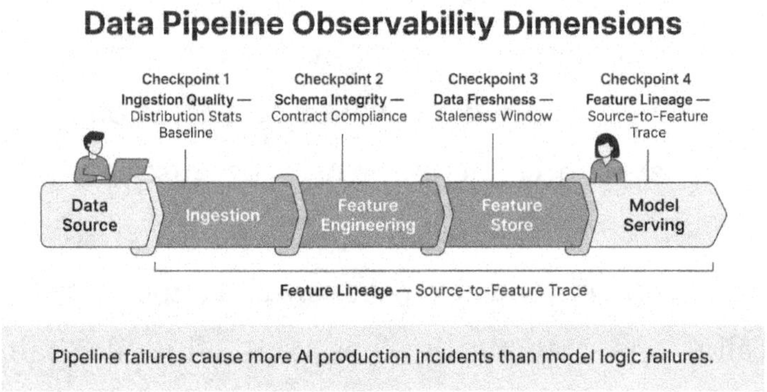

Five-layer AI observability stack: signal types, failure modes, and diagnostic questions per layer

5.3 Infrastructure and Data Pipeline Layers: GPU Telemetry, Queue Latency, and Feature Integrity

The bottom two layers are the foundation on which everything else depends. Layer 1 covers the compute environment: GPU clusters, inference servers, networking, and the queues that buffer requests between clients and model serving workers. Layer 2 covers the data supply chain: ingestion jobs, transformation pipelines, feature stores, schema registries, and the freshness SLOs governing how stale a feature vector is allowed to be. Both layers reuse traditional APM patterns but require AI-specific instrumentation to emit the signals that higher layers consume.

GPU Telemetry and the USE Method Adapted for ML

The USE method (Utilization, Saturation, Errors) maps onto GPU infrastructure when applied with ML-specific precision. Utilization means SM (streaming multiprocessor) utilization, not the coarse "GPU busy" percentage that cloud dashboards default to. An inference server can report 70 percent GPU utilization while actual SM utilization is 40 percent because memory bandwidth is the bottleneck. The NVIDIA DCGM exporter exposes per-SM utilization, memory bandwidth, and NVLink throughput metrics; wire those into your OpenTelemetry pipeline and configure the OTel Collector to forward them with the gen_ai.request.model and ml.model.version span attributes.

Queue depth is the leading indicator of latency SLO violations. Alert on queue depth rather than on latency: by the time p99 crosses your SLO threshold, users have already experienced the degradation. Instrument the inference server (vLLM, TorchServe, Triton) to emit queue depth as an OpenTelemetry gauge metric with span attributes for model name, version, and serving replica ID. GPU memory pressure manifests as KV cache eviction rates: when the KV cache fills, generation latency increases non-linearly. A drop in KV cache hit rate from 85 percent to below 60 percent signals either a shift in request length distribution or a new model version with a larger memory footprint than the serving configuration was provisioned for.

Decompose p99 inference time into child spans: preprocessing, queue wait, GPU forward pass, and post-processing. Use OpenTelemetry GenAI semantic conventions for core span attributes (gen_ai.system, gen_ai.request.model, gen_ai.usage.input_tokens, gen_ai.usage.output_tokens) and add custom attributes for queue wait milliseconds, actual batch size, and KV cache hit rate. The span decomposition tells an on-call engineer within thirty seconds whether the bottleneck is queue wait, GPU compute, or post-processing.

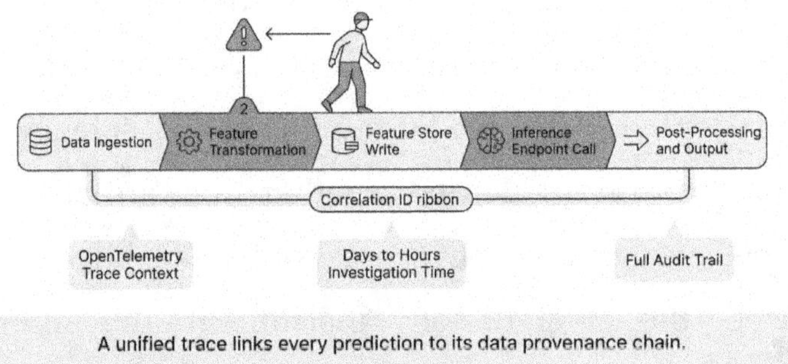

OpenTelemetry span decomposition for LLM inference: queue wait, GPU forward pass, KV cache hit rate, post-processing child spans

Data Pipeline Observability: Freshness SLOs, Schema Validation, and Feature Lineage

Feature pipeline observability begins with a freshness SLO per feature group: a real-time fraud model might allow 90 seconds of staleness; a nightly-updated credit model might allow 26 hours. Store SLOs as machine-readable metadata in the feature store and continuously evaluate them. Instrument the feature store to emit a feature.freshness_seconds metric per feature group and alert when freshness exceeds the SLO threshold for any group with an active model consumer. Schema validation should occur before feature computation: register every expected input schema in a schema registry and configure each ingestion job to validate incoming records, emitting a validation span with the failure rate and failure categories. A spike in upstream schema failures is a leading indicator of feature drift to catch before it manifests in production.

Data quality SLIs extend beyond schema correctness into statistical consistency: null rate, out-of-range rate, and distribution stability score computed via Population Stability Index against a rolling reference window, exported as OpenTelemetry metrics per feature group. Data contracts formalize this interface as machine-readable specifications declaring the schema, freshness SLO, statistical SLIs, and the on-call owner, stored in the schema registry and enforced at read time. Feature lineage closes the loop by propagating a trace correlation ID from the ingestion job through every transformation stage to the feature store write, so that any feature value can be traced to its raw source record in a single query.

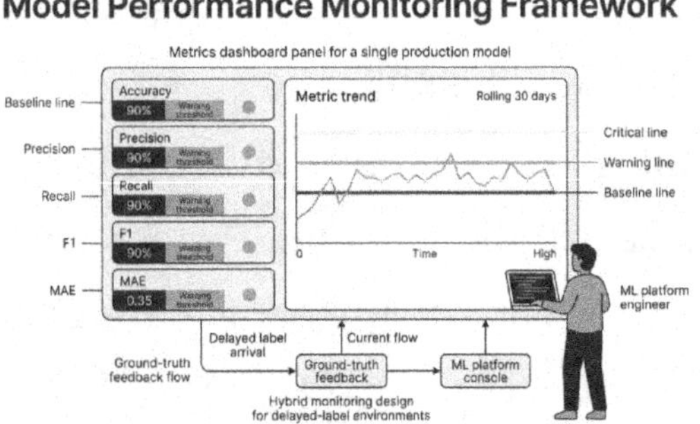

Each production model requires a documented monitoring specification: baseline, warning, and critical thresholds.

Data contract enforcement: schema registry, freshness SLO evaluation, and distribution SLI monitoring per feature group

5.4 Model Behavior Layer: Output Distribution, Confidence Calibration, and Slice Metrics

Layer 3 is where AI-specific monitoring begins in earnest. Infrastructure and data pipeline monitoring reuses traditional APM patterns. Model behavior monitoring requires instrumentation primitives with no direct analog in traditional software observability: output distribution tracking, prediction confidence calibration, and performance disaggregation across population subgroups. These signals indicate whether the model you deployed is still the one you validated.

Output Distribution Monitoring and the RED-Plus-Distribution Framework

The RED method (Rate, Errors, Duration) needs a fourth dimension for ML: Distribution. Rate, Errors, and Duration describe the serving infrastructure. Distribution describes whether the statistical character of the model's outputs has changed relative to a reference window. Without it, RED-based monitoring conceals the most common class of model failures. Instrument your model serving layer to emit output distribution histograms using OpenTelemetry Histogram instruments with explicit bucket boundaries matched to the model's output range. Compute the Population Stability Index (PSI) and the Kolmogorov-Smirnov statistic over a rolling reference window at each aggregation interval. PSI below 0.1 is stable; between 0.1 and 0.2 triggers a warning; above 0.2 triggers an alert and opens an incident ticket. Use a five-minute aggregation interval as the default; reduce to one minute for high-stakes endpoints where rapid drift detection justifies the storage cost.

For LLMs, track response length distribution, the fraction of responses hitting maximum context length, and the distribution of finish reasons (stop, length, content_filter) as categorical counters. A rise in content_filter is a safety signal; a rise in length is a quality signal.

LLM Output Evaluation Pipeline

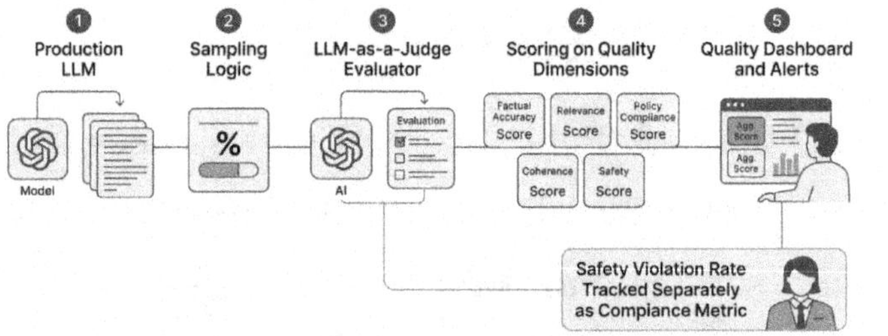

LLM quality requires a semantic evaluation pipeline — statistics alone are insufficient.

Output distribution monitoring: PSI time series, histogram overlay current vs. reference window, finish-reason counter distribution for LLM endpoints

Prediction Confidence Calibration and Slice-Based Metrics

Confidence calibration is one of the most frequently unmonitored properties of models in production ML. A well-calibrated model produces confidence scores that match empirical accuracy. When the model says it is 80 percent confident, it should be correct about 80 percent of the time within that confidence band. An uncalibrated model claiming 90 percent confidence while achieving only 60 percent accuracy undermines every downstream decision that relies on that score. Instrument your model serving layer to emit the raw confidence score for every prediction as a span attribute (ml.model.output.confidence). Aggregate scores into buckets, track empirical accuracy within each bucket as ground-truth labels arrive, compute Expected Calibration Error (ECE) per bucket, and export it as a gauge metric—Alert when ECE exceeds 0.05 for models used in automated decision-making. A shift in confidence distribution is also a leading indicator of model drift: a bimodal distribution shifting toward uniformity correlates with concept drift even before accuracy metrics degrade.

Aggregate accuracy metrics conceal failure modes visible only in specific subgroups. A credit model with 92 percent overall accuracy may have 78 percent accuracy for applicants in a specific geographic region. Define performance slices at model deployment time as predicates over the input feature space (geographic region, customer segment, product category, user language) and track the same performance metrics at the slice level, exported via OpenTelemetry with ml.model.slice.name and ml.model.slice.predicate span attributes. Tier slices by monitoring frequency: tier-1 runs continuously; tier-2 evaluates daily; tier-3 evaluates weekly. Route tier-1 slice alerts through the same on-call pipeline as aggregate drift alerts.

Agentic AI Observability Components

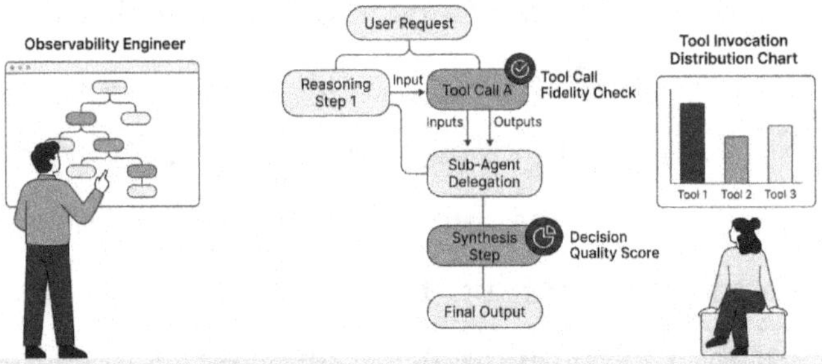

Agent observability requires decision graph capture — linear traces are insufficient.

Slice-based performance monitoring: per-segment accuracy time series with SLO threshold bands overlaid on aggregate metric

5.5 Evaluation and Governance Layers: Online Eval, Golden Test Sets, Audit Log Architecture, and Attestation Pipelines

Layers 4 and 5 address the questions that statistical monitoring cannot answer. Layer 4 asks whether individual model outputs are correct, safe, and policy-compliant right now, at the semantic level. Layer 5 asks whether there is a verifiable, tamper-evident record proving what was deployed, when, by whom, with what approval, and what the eval scores were at every audit interval. These are the layers where most ML platforms under-invest, and where the consequences are most severe during a regulatory review.

Online Evaluation: Evaluator-as-Judge and Golden Test Sets

Online evaluation runs quality assessments on live production traffic rather than on held-out validation sets, which reflect the distribution at training time rather than the current one. The evaluator-as-judge architecture deploys a second model to score the primary model's outputs on a continuous sample, returning a structured score covering factual grounding, policy compliance, relevance, response completeness, and safety. For safety-critical dimensions, run the evaluator on 100 percent of production traffic; for quality dimensions where cost matters, run it on a representative sample (typically 5 to 10 percent, stratified by user segment). Tools like LangSmith, Langfuse, and Arize Phoenix provide evaluator pipeline infrastructure, including prompt management, sampling logic, and scoring aggregation at production scale.

Instrument the evaluator pipeline as a child span of the primary inference span, carrying span attributes for each rubric dimension score and the sampling decision. When an eval score drops below the SLO threshold, the alert links directly to the specific trace, allowing the on-call engineer to inspect the input, output, and rubric score in a single navigation step. Golden test sets are the offline complement: curated input-output pairs with human-verified correct outputs. Run the golden test set against every candidate model version before promotion; a regression relative to the deployment baseline is mandatory. Automate the evaluation in your CI/CD pipeline: pull the test set from versioned storage, compute per-dimension scores, compare against baseline thresholds in the model registry, and emit a structured evaluation report as a signed registry artifact.

OpenTelemetry AI Stack Integration

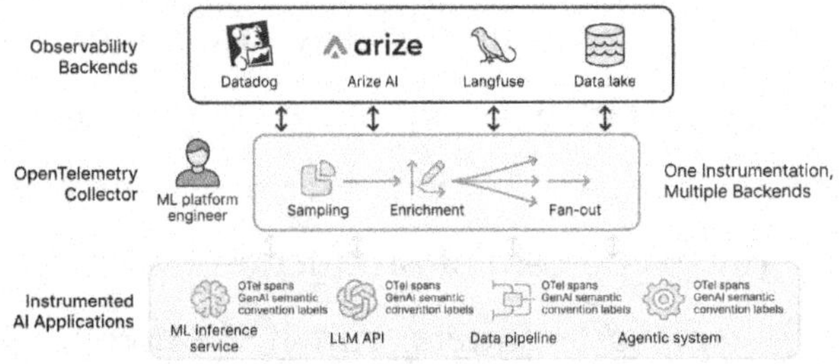

OpenTelemetry provides a vendor-neutral telemetry language for the full AI observability stack.

Online eval pipeline: evaluator-as-judge span as child of inference span, rubric dimension scores, golden test set regression gate in CI/CD

Audit Log Architecture and the Attestation Pipeline

In the AI observability context, an audit log is a structured, append-only event store with cryptographic integrity guarantees that covers every material model lifecycle event. For each event, record the timestamp (UTC, millisecond precision), the event type, the subject (model name, model version hash), the actor, the authorization record, the outcome, and the trace correlation ID from any associated inference request. Write the audit log to both an immutable event store (Apache Kafka with log compaction disabled, or AWS QLDB for stronger cryptographic guarantees) and to a structured data lake table, partitioned by date and model name, for queryability.

Model registry signing ties a specific trained artifact to a specific set of approval decisions: the pipeline computes a SHA-256 hash of the artifact, records it in the registry, and signs the registry record with the platform team's signing key. Any subsequent deployment checks the artifact hash; a mismatch blocks the deployment and logs an alert error. The attestation pipeline extends signing to cover the full lifecycle: an attestation is a signed, structured record asserting that a specific model version passed a specific check. The golden test set pipeline issues one when a model version passes the eval gate; the model risk team issues one when a model version is cleared for production. Wire attestation checks are integrated into deployment automation so that a deployment with a missing or expired attestation fails with a structured error that references the missing attestation type. Instrument attestation issuances as OpenTelemetry spans with attributes for model name, model version hash, attestation type, issuer identity, and expiration timestamp.

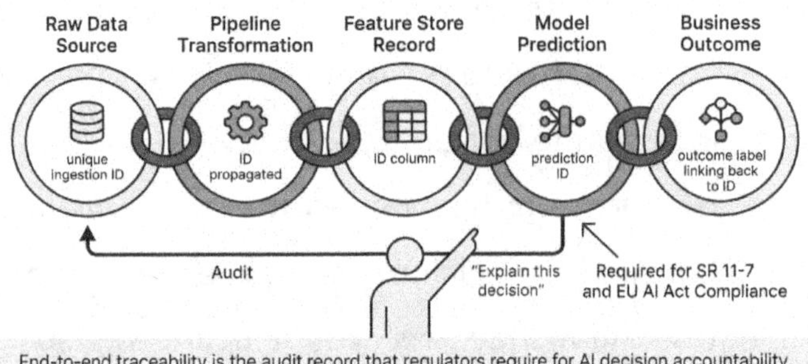

End-to-end traceability is the audit record that regulators require for AI decision accountability.

Attestation pipeline: model artifact hash, signed attestation chain, deployment gate checks, and audit log write per lifecycle event

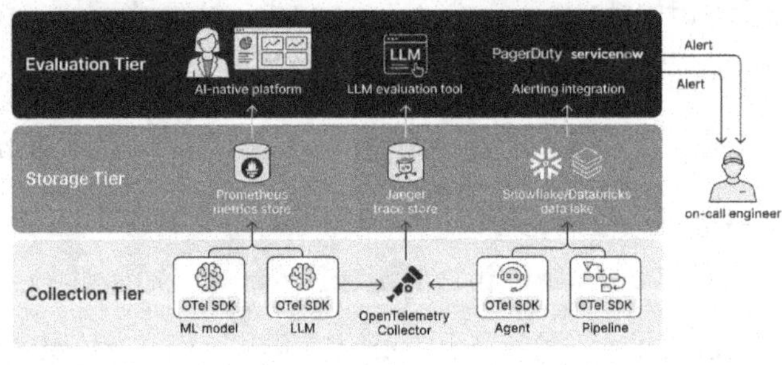

The AI observability reference architecture delivers complete coverage only when all three tiers are connected.

End-to-end trace: ingestion correlation ID propagated through feature store, inference span, evaluator child span, and attestation log record

5.6 Technical Team Checklist: Five-Layer Instrumentation Plan

- **Action:** Instrument the GPU inference cluster with the NVIDIA DCGM exporter and configure the

OpenTelemetry Collector to forward SM utilization, memory bandwidth, KV cache hit rate, and inference queue depth as metrics with model name and version labels. **Owner:** ML Platform Eng. **Done when:** SM utilization, queue depth, and p99 inference latency are queryable in a single dashboard panel with five-minute resolution and documented SLO thresholds.

- **Action:** Define freshness SLOs for every feature group with an active model consumer, store them in the feature store metadata, and configure continuous freshness evaluation alerts. **Owner:** Data Eng. **Done when:** Any feature group exceeding its freshness SLO triggers an alert within two scrape intervals, identifying the model consumer and observed staleness in seconds.
- **Action:** Register all expected input schemas in the schema registry and add schema validation spans to every ingestion job, emitting validation failure rates and failure categories as OpenTelemetry histogram metrics. **Owner:** Data Eng. **Done when:** Schema validation failure rate is visible in the central observability dashboard, and any spike above 0.5 percent triggers an alert.
- **Action:** Instrument model serving endpoints to emit output distribution histograms at five-minute resolution, compute PSI and KS-statistic against a 30-day reference window, and alert when PSI exceeds 0.2. **Owner:** ML Platform Eng. **Done when:** PSI and KS-statistic time

series are queryable alongside latency and error rate, and a PSI alert has been validated with a synthetic drift injection test.

- **Action:** Deploy an evaluator-as-judge pipeline sampling 10 percent of production LLM traffic, scoring each sample on factual grounding, policy compliance, and safety rubric dimensions, with eval scores emitted as child spans of the primary inference span. **Owner:** ML Engineer. **Done when:** Rubric dimension scores are visible in the trace viewer, and a score below the SLO threshold triggers an on-call alert within five minutes linking to the offending trace.
- **Action:** Build and maintain a golden test set of at least 200 input-output pairs with human-verified outputs, integrate evaluation into the model promotion CI/CD pipeline as a mandatory gate, and store evaluation reports in the model registry as signed artifacts. **Owner:** ML Engineer. **Done when:** No model version can reach production without a linked passing evaluation report, and any regression below the baseline triggers an automatic promotion block.
- **Action:** Implement an append-only audit log capturing all model lifecycle events with trace correlation IDs, written to both an immutable event store and a queryable data lake table partitioned by date and model name. **Owner:** Governance Eng. **Done when:** A compliance

engineer can answer which model version was serving on a given date and what its eval scores were in a single SQL query.

- **Action:** Build a model registry signing pipeline computing SHA-256 hashes of model artifacts at registration and verifying artifact integrity before every deployment, blocking mismatches with a logged, alerted error, and define an attestation requirement policy with deployment gate enforcement per model risk tier. **Owner:** ML Platform Eng / Governance Eng. **Done when:** A deployment attempt with a missing attestation or a hash mismatch fails with a structured error that references the specific violation, and a deliberate gate-bypass test validates the failure.

5.7 Team Conversation Starters

- If every layer of the stack fired an alert simultaneously tonight, which one would the on-call engineer investigate first, and does your current runbook explicitly answer that prioritization question?
- For each feature group feeding a production model, can your team answer in under five minutes whether a freshness SLO breach occurred in the last 24 hours and whether it correlated with any observable shift in model output distribution?

- When the evaluator-as-judge scores a production LLM response as a policy violation, what is the exact sequence of automated actions and human escalations that follows, and what is the documented SLO for time-to-remediation?
- Can you trace any production inference request from the raw ingestion record through the feature store to the final output and the associated eval score in a single observability session without leaving the trace viewer?
- If a regulator asked today for the full decision audit trail for a specific model prediction made six months ago, including the model version, the feature values, the active attestations, and any eval alerts that fired within 24 hours, how long would it take your team to produce that record?
- Does your model promotion pipeline enforce that a golden test set evaluation report and all required attestations are present before any model version reaches production, and has that gate been validated with a deliberate bypass attempt?

5.8 Key Takeaway

The five-layer AI observability stack is the minimum viable instrumentation architecture for a production AI system you can operate under load, debug during an incident, and audit under regulatory scrutiny. Each layer answers a distinct diagnostic question that the other four cannot: infrastructure health, data pipeline integrity, model behavioral stability, semantic output quality, and governance provenance. Instrumenting fewer than five layers converts saved instrumentation time into larger debugging time during the incidents that incomplete coverage allows to run longer. The architectural principle that connects all five layers is the shared trace context: OpenTelemetry correlation IDs are propagated from data ingestion through feature engineering, model inference, output evaluation, and attestation logging. With that thread in place, the five layers become a single navigable observability plane, and the gap between alert and root cause shrinks from hours to minutes.

6 ML Model Monitoring: Keeping Traditional Models Production-Ready

6.1 Opening Scenario

Maya Krishnamurthy had been running the fraud detection classifier in production for fourteen months without a P1 incident. The gradient-boosted model scored every card-present transaction in under 30 milliseconds, the latency dashboard was green, and the weekly model health email showed AUC holding at 0.917. Then, on a Tuesday standup, the fraud operations team mentioned their manual review queue had ballooned. Chargebacks were up 19 percent over a 30-day window. Maya pulled the raw score distribution and saw that the shape was identical to baseline. Accuracy sat at 99.1 percent.

She almost closed the dashboard. Instead, she ran a slice-based breakdown she had wired three weeks earlier: AUC segmented by merchant category code. The aggregate number was masking a disaster. One high-volume grocery chain, accounting for roughly 8 percent of daily transaction volume, had a slice AUC of 0.61. The model was barely better than random for that merchant. The chain had migrated its point-of-sale terminals to a new processor the previous month, thereby shifting the distribution of transaction amounts and introducing a new terminal ID encoding. The feature store was still serving the old encoding, and the training-serving skew had rotted the model's decision boundary for that entire segment.

The fix took four hours: a targeted feature pipeline patch, a slice-aware retrain on the affected merchant cohort, and a shadow-scored canary deployment to validate performance before promoting. The detection had taken three weeks of silent degradation because the monitoring program measured aggregate accuracy rather than segment-level AUC. After the incident, Maya instrumented slice dashboards for every high-volume merchant and wired PSI alerts on the terminal ID feature to fire within one monitoring cycle. The aggregate number never told the story. The slice did.

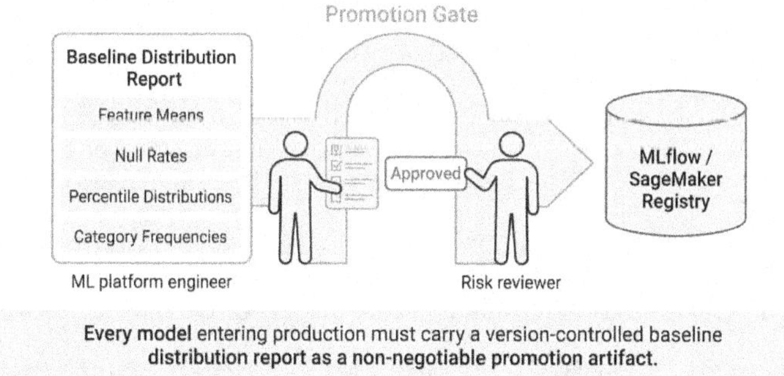

Slice-level AUC breakdown exposing a high-volume merchant segment near random performance, while aggregate AUC holds steady

6.2 Why It Matters for Production Systems

Classical ML monitoring is its own engineering discipline, distinct from infrastructure observability and distinct from LLM evaluation practice. A latency dashboard confirms that inference endpoints are responding. A feature pipeline health check confirms ETL jobs completed. Neither signal tells you whether the model is making correct decisions, whether confidence scores are calibrated to reality, or whether a high-impact segment has drifted outside the support of the training distribution. Infrastructure uptime and model correctness are orthogonal. A model can be simultaneously operational and wrong.

The core challenge is label arrival lag. In online fraud detection, labels arrive within hours because chargebacks close the loop quickly. In credit underwriting, the ground-truth label may not materialize for 12 to 36 months after scoring. In clinical risk stratification, outcome labels can lag inference by years. This latency means the standard offline evaluation loop cannot run at inference cadence. The monitoring program must operate in two parallel modes: a fast loop that monitors input and output distributions without labels, and a slow loop that joins model predictions to arriving ground truth via a label arrival pipeline and computes lagged performance metrics.

The label arrival pipeline is a first-class data engineering system. It must ingest outcome events from downstream systems, perform a ground-truth join against logged inference records using a stable prediction ID, handle late arrivals and corrections, and write labeled evaluation windows into the metrics backend at configurable cadences. Building it correctly requires the ML platform engineer who owns the inference logger, the data engineer who owns the outcome event stream, and the ML engineer who defines the labeling logic to work from a shared schema. Missing a late-arriving label causes the monitoring program to compute performance metrics on a biased sample of early-resolving cases, a systematic error that compounds over time.

Inverse propensity weighting adds complexity wherever the model's own decisions determine which outcomes are observable. A credit model rejecting 40 percent of applicants can never observe the ground-truth default rate for that rejected population. The performance metrics computed on the approved-and-observed population are biased estimates of true model performance across the full applicant pool. Correcting this requires reweighting each observed outcome by the inverse of its estimated probability. Most off-the-shelf monitoring platforms do not implement this out of the box; if the model's decisions introduce selection bias in the label population, the ML platform team must explicitly build or integrate the correction.

Training-serving skew is the other major classical failure mode. It occurs when the feature values a model receives at inference time differ systematically from those seen during training. The causes are numerous: diverging preprocessing logic between training and serving pipelines, different versions of categorical encoders, time-zone handling mismatches, null imputation strategies that vary between offline and online paths, and feature store configurations that inconsistently apply transformations. Training-serving skew does not trigger infrastructure alerts because the model server receives valid requests and returns valid responses. Degradation appears only in prediction quality; it is invisible without a monitoring program that explicitly compares features arriving at the serving endpoint with those captured during training.

The regulatory dimension is concrete. SR 11-7 from the Federal Reserve requires ongoing performance monitoring, independent validation, and documented retraining triggers at regulated financial institutions. The EU AI Act classifies credit scoring and loan origination models as high-risk AI systems subject to ongoing monitoring obligations that must be documented and auditable. The ML platform engineer who owns the monitoring infrastructure also makes the organization's compliance posture defensible, and that ownership stake is serious.

6.3 Performance Metric Monitoring: Online and Offline Signal Alignment

A production ML model requires two complementary performance monitoring tracks. The offline track computes metrics against ground-truth labels as they arrive through the label arrival pipeline. The online track monitors proxy signals, input distributions, output score distributions, calibration error, and threshold-crossing rates, providing leading indicators of degradation before labels are available. Neither is sufficient alone. Offline metrics are accurate but lagged. Online metrics are immediate but indirect. Both tracks should write to the same metrics backend, with on-call alert routing configured to distinguish between them.

Accuracy is the wrong primary metric for any imbalanced-class problem, which describes most fraud, churn, and clinical risk classifiers. The monitoring dashboard should expose precision, recall, and F1 at the operational threshold alongside AUC-ROC for threshold-independent ranking quality and AUC-PR for positive-class tail performance. Log loss complements AUC by penalizing confident wrong predictions more heavily than uncertain wrong ones: a model whose AUC holds steady but whose log loss increases is losing calibration over time. For models where raw output scores drive downstream decisions rather than a hard threshold, calibration drift is particularly dangerous. Monitor expected calibration error (ECE) as a first-class metric, visualized as a reliability diagram showing predicted probability bins against observed frequencies, and wire an alert when ECE exceeds the threshold established during the burn-in period.

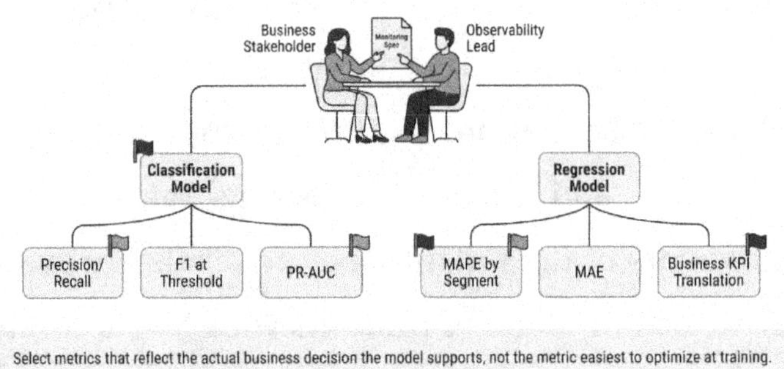

Reliability diagram showing calibration drift: predicted probability bins versus observed frequencies drifting above the diagonal over a 90-day production window

Slice-based metrics surface the failure mode from Maya's scenario. Aggregate metrics can be stable while a high-impact segment deteriorates. Compute performance metrics disaggregated by every dimension the model's decisions affect differentially: merchant category, geography, customer tenure band, channel, and any protected attribute required by fairness regulations. Segment dashboards should be first-class views in the observability platform, not buried in quarterly reports. Wire alerts on segment-level AUC drops with the same urgency as alerts for the aggregate metric. A segment-level collapse is typically the leading indicator of what will appear in aggregate metrics two to four weeks later.

For the online track, prediction confidence histograms are the most immediately useful signal. Plot the distribution of raw model output scores over a sliding 24-hour window and overlay against the baseline distribution captured at promotion time. A bimodal distribution collapsing toward a single mode, or a mean shift exceeding one standard deviation, indicates the model's decision boundary has moved without waiting for labels. This requires only that the inference logger captures every raw score, a non-negotiable instrumentation requirement at deployment. The inference log is the foundation of every downstream monitoring capability.

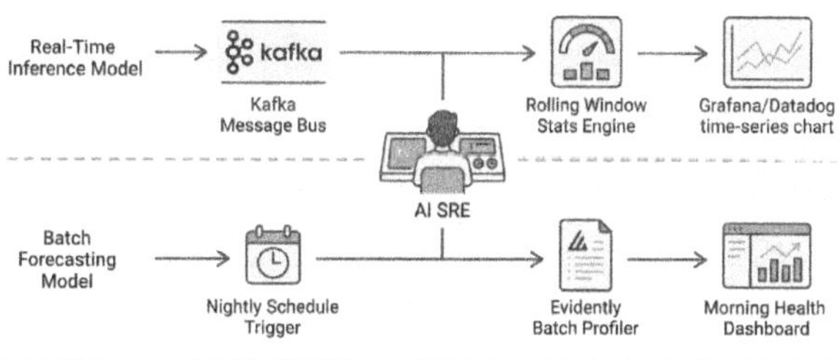

Online prediction confidence histogram: 30-day sliding window overlaid against promotion-time baseline, showing mean shift indicating training-serving skew

6.4 Input and Output Distribution Monitoring: Drift Detection at Feature Resolution

Distribution monitoring provides the fastest feedback loop for detecting model degradation before ground-truth labels arrive. A model trained on a specific joint feature distribution produces reliable predictions only while production inputs remain within that distribution's support. Detecting drift early, at the feature and prediction output levels, is the primary function of the online monitoring track. Three statistical methods form the core of this detection layer.

The Population Stability Index (PSI) is the standard metric for measuring continuous feature drift in financial services and related domains. PSI computes a symmetric KL-divergence-like statistic across binned reference and production distributions. The conventional thresholds are: below 0.1 for stable distribution, 0.1 to 0.2 for moderate drift requiring investigation, and above 0.2 for significant drift triggering a retraining evaluation. These are starting points to calibrate against the model's sensitivity, each feature's importance rank, and the cost of a false alarm versus a missed drift event. The Kolmogorov-Smirnov test provides a complementary signal, particularly useful for categorical features where PSI's binning approach can misrepresent sparse categories. The KS test compares empirical CDFs and returns a p-value indicating whether the reference and production samples could plausibly share the same underlying distribution; it is distribution-free and computationally inexpensive, making it practical for monitoring hundreds of features in a streaming context.

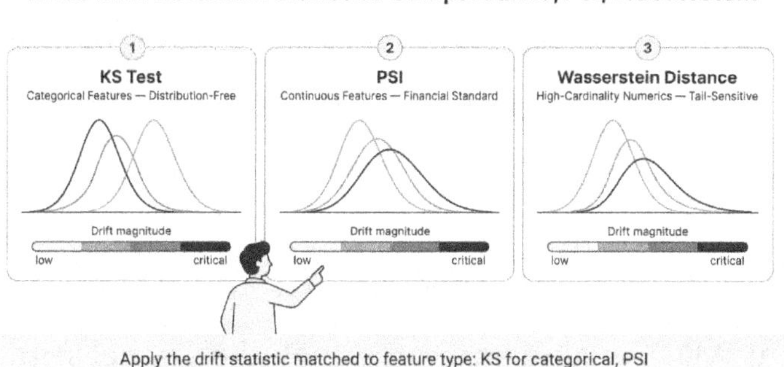

PSI heatmap across 47 input features over 90 days, with the terminal ID encoding feature crossing the 0.2 threshold in week 6 before segment AUC collapse

Wasserstein distance addresses the KS test's limitation of weighting all distribution regions equally. It computes the minimum transport cost between two distributions, sensitive to both the magnitude and location of shifts. For high-cardinality numerical features such as transaction amounts and recency-frequency-monetary features, Wasserstein distance provides a more nuanced drift signal. For features identified as high-importance during model development, the additional computational cost is justified by the additional diagnostic resolution. Feature-level drift alerting must be tiered by feature importance: a PSI of 0.25 on a top-3 feature routes to the critical channel and pages the on-call

ML engineer within minutes, while the same PSI on a feature ranked 40th routes to a Slack channel for async review. This routing logic belongs in the observability platform configuration, not an informal team norm.

For models that use tabular embeddings as intermediate representations, per-feature PSI does not apply to the learned vector representation. Monitor embedding drift using Maximum Mean Discrepancy (MMD), which estimates the distance between two distributions in a high-dimensional space using kernel methods. A statistically significant MMD value indicates that the model's internal representation of the input space has shifted, even when no individual input feature crosses a threshold.

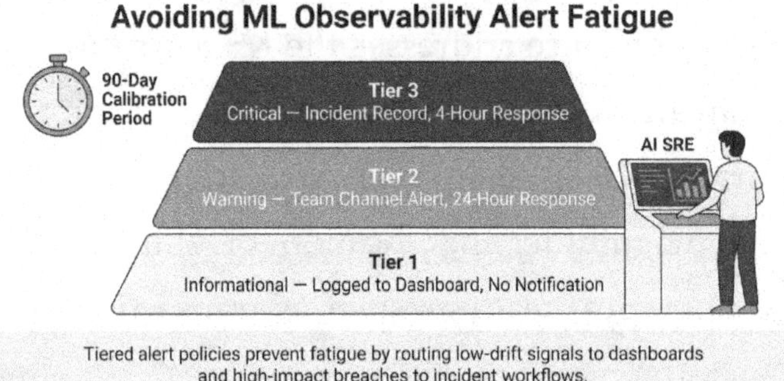

Tiered alert policies prevent fatigue by routing low-drift signals to dashboards and high-impact breaches to incident workflows.

MMD embedding drift signal for a customer behavioral embedding, showing a gradual shift over 8 weeks with acceleration following a product launch

Segment-level alerting extends feature drift monitoring to business-meaningful subpopulations. Configure the monitoring platform to compute PSI and KS statistics separately for each major segment: geography, customer cohort, product line, and channel. A drift event diluted to PSI 0.08 in aggregate may be driven by PSI 0.31 in a geography accounting for 12 percent of volume. Segment-level drift that precedes segment-level performance degradation is the earliest detectable signal in the monitoring stack. Configure it at model deployment time, not as a post-incident enhancement. Output distribution monitoring closes the loop: wire the inference logger to capture the full raw score or probability vector, compute the distribution over a rolling window, and compare it against the baseline using the same statistical tests applied to the input features. A prediction distribution that shifts without any corresponding input feature drift suggests a non-linear interaction effect that per-feature monitors cannot capture, a common failure mode in tree ensemble models.

6.5 Operational Monitoring and Retraining Triggers: Closing the Feedback Loop

Performance and distribution monitoring answer whether the model is making good decisions. Operational monitoring answers whether the infrastructure can deliver those decisions at the required scale and latency. Both must be instrumented in a production monitoring program. Instrument the model server to emit p50, p95, and p99 latency for every inference request, tagged by model version and preprocessing step. For models that rely on online feature lookups from a feature store such as Feast or Tecton, instrument the feature-store round-trip as a separate span within the inference trace and export it via OpenTelemetry to the central observability backend. Without span-level attribution, latency spikes caused by feature store capacity and by model scoring complexity look identical in aggregate metrics. Wire queue depth (for batch systems) and consumer lag (for streaming systems) into the same SLO dashboard as latency and throughput, with alerts firing before queue growth causes downstream timeouts.

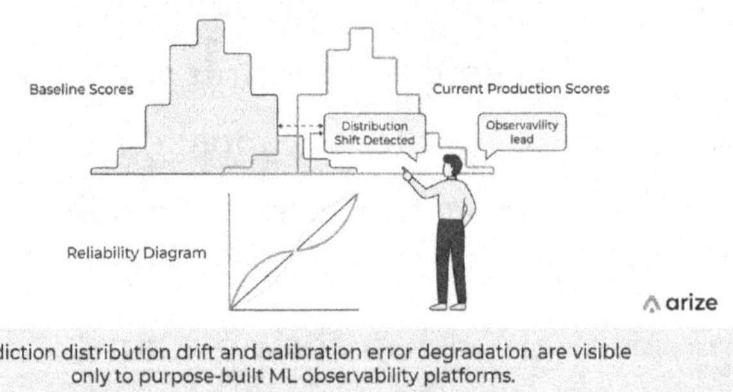

Prediction distribution drift and calibration error degradation are visible only to purpose-built ML observability platforms.

Inference trace with span decomposition: feature store lookup, preprocessing, model scoring, and post-processing, showing feature store as the dominant p99 latency contributor

Fallback paths must bc instrumented with the same rigor as primary paths. Instrument every fallback activation with a reason code (timeout, error, feature unavailability) and route those codes to a dedicated monitoring channel. A fallback rate above the configured threshold (typically 1 to 5 percent of requests) indicates a systemic problem in the primary path. Shadow scoring is the safest way to evaluate a candidate model against live production traffic. In shadow mode, the challenger model receives a copy of every inference request, scores it, and logs the result without its output ever reaching the caller. Wire the shadow scorer into the same inference logger as the production model, sharing a common prediction ID so champion and challenger results can be joined for direct comparison. This provides a realistic traffic sample that an offline held-out evaluation cannot replicate.

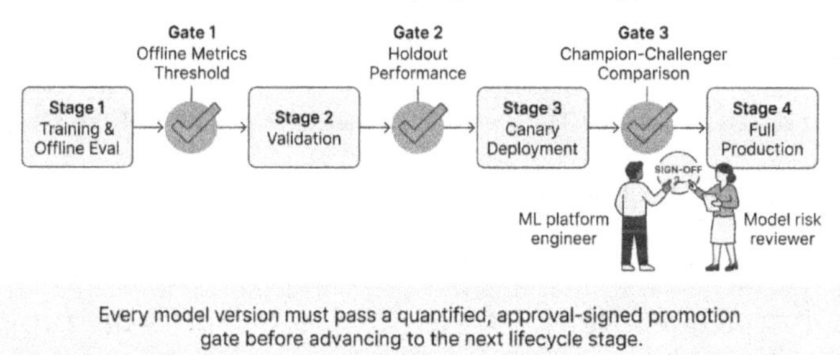

Shadow scoring architecture: champion serving production traffic while challenger receives mirrored requests, both logging to a common inference store for head-to-head comparison

Blue/green model rollouts provide safe promotion mechanics once shadow scoring validates the challenger. Traffic routing stays pointed at the blue (incumbent)

environment after the challenger is promoted to a live green environment. Traffic shifts in step-wise progression: 5, 25, 50, then 100 percent. Each step is gated by an automated check of segment- and distribution-level metrics; if any check fails, traffic is routed back to the blue environment without manual intervention. The blue environment remains live and warm throughout, so rollback is a configuration change with no cold-start latency.

Retraining triggers should be codified as explicit decision logic rather than organizational convention. Maintain a state machine for each production model: nominal (all metrics within thresholds), degraded-watch (one or more metrics outside thresholds under investigation), degraded-retrain-recommended (sustained degradation above threshold with quantified business impact), and retrain-in-progress (retrain pipeline running, shadow scorer active). Transition from degraded-watch to retrain-recommended requires sustained drift across multiple monitoring windows, plus a confirmed business-impact estimate from the label-arrival pipeline. Every production model also needs a kill switch: a configuration flag that routes all traffic to the fallback path instantly, without a code deploy, and is operable by the on-call SRE without ML engineer involvement. Test the kill switch in a scheduled game-day exercise at least quarterly and document the procedure in the model's runbook.

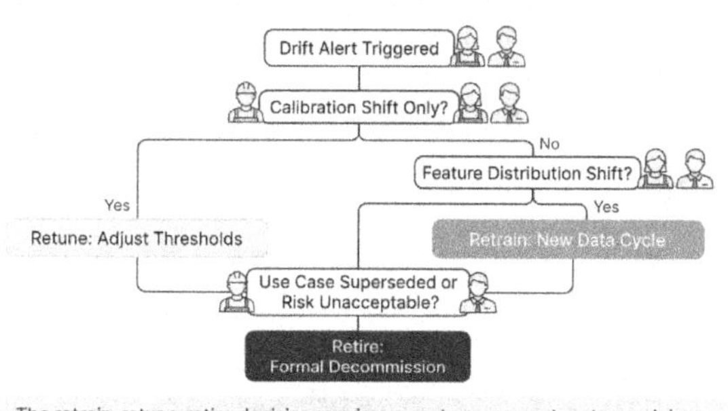

The retrain-retune-retire decision requires a cost assessment, not an opinion: confirm drift type before committting to the most expensive option.

Model serving state machine: nominal, degraded-watch, and retrain-in-progress states with automated transition gates and kill-switch fallback routing

6.6 Technical Team Checklist: Tabular and Classical ML Monitoring

- **Action:** Generate a baseline distribution report at promotion time covering mean, standard deviation, percentile distribution, null rate, and category frequency for every input feature, stored as a versioned artifact in the model registry alongside the model binary. **Owner:** ML Platform Eng. **Done when:** Every promoted model has a linked baseline artifact queryable via the registry API, and the promotion gate rejects models without one.
- **Action:** Implement a label arrival pipeline that joins inference logs to outcome events using a stable prediction ID, handles late arrivals up to the configured maximum label delay, and writes labeled evaluation

windows to the metrics backend at the configured cadence. **Owner:** Data Eng. **Done when:** Offline performance metrics (AUC, precision, recall, ECE) are auto-computed for each label arrival batch and visible in the model's monitoring dashboard within two hours of batch completion.

- **Action:** Instrument the model server with per-request span-level telemetry covering feature store lookup, preprocessing, model forward pass, and post-processing, tagged by model version and request source, exported via OpenTelemetry to the central metrics backend. **Owner:** ML Platform Eng. **Done when:** p50, p95, and p99 latency are queryable per-span at 1-minute resolution for every production model version.
- **Action:** Configure PSI alerts for every continuous input feature, KS alerts for every categorical feature, and MMD alerts for every tabular embedding, with alert severity weighted by feature importance rank and routed appropriately: page for top-5 features, async channel for remaining. **Owner:** ML Engineer. **Done when:** A synthetic drift injection test triggers the correct alert tier for each feature class within one monitoring cycle.
- **Action:** Set up segment-level AUC and prediction distribution dashboards disaggregated by each high-volume business segment, with alert thresholds calibrated during a 90-day burn-in period of shadow-

mode monitoring. **Owner:** ML Engineer. **Done when:** Segment dashboards are linked from the model's registry entry, and alert thresholds are documented in the monitoring specification.

- **Action:** Deploy every candidate model into shadow scoring mode against full production traffic before canary promotion, capturing challenger scores in the same inference logger as the champion model with a shared prediction ID for join analysis. **Owner:** ML Platform Eng. **Done when:** Shadow scoring is a mandatory gate in the CI/CD pipeline, and challenger-vs-champion comparison reports are auto-generated after each shadow window.
- **Action:** Implement a blue/green traffic-shifting rollout with automated step gates at 5, 25, 50, and 100 percent that check segment-level and distribution-level metrics at each step and roll back automatically on gate failure. **Owner:** ML Platform Eng. **Done when:** A rollback is demonstrated successfully in a game-day exercise with gate check logic verified against a synthetic degradation injection.
- **Action:** Document and test the kill-switch procedure for every production model so the on-call SRE can route all traffic to the fallback path within 5 minutes without a code deploy, and run a kill-switch game day at least quarterly. **Owner:** SRE / On-Call. **Done when:** The kill-

switch runbook is in the on-call handbook, the last game-day result is documented, and the next scheduled game day is on the team calendar.

- **Action:** Codify the retraining trigger state machine for each production model with explicit criteria for the transition from degraded-watch to retrain-recommended, including sustained-drift duration, business impact threshold, and approver chain. **Owner:** ML Engineer. **Done when:** The state machine is implemented as a configuration artifact in the model registry, and a simulated trigger test verifies that correct notifications are emitted at each state transition.

Cross-Environment Observability Continuity for ML Models

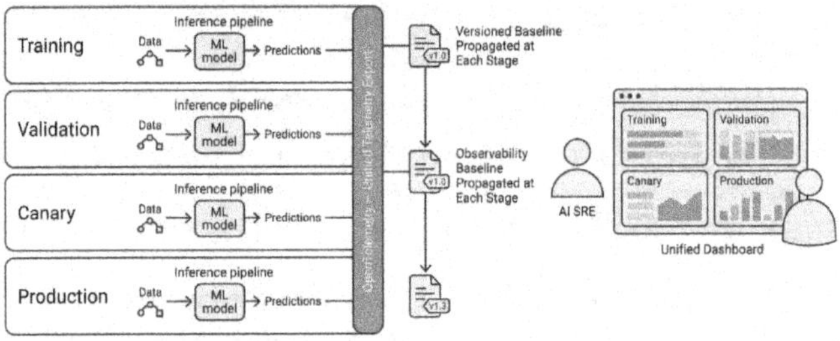

Observability instrumented from training through production detects environment discrepancies before they become production incidents.

Model lifecycle governance flow: promotion gate, shadow scoring, blue/green rollout, and retraining state machine with documented ownership at each decision point

6.7 Team Conversation Starters

- When the aggregate AUC held steady last quarter, which segments did we not check, and how quickly can we get segment-level performance metrics for those cohorts right now?

- What is the current label arrival delay for each production model, and which models have delays long enough that our online distribution monitoring is the only real-time degradation signal we have?
- If we needed to activate the kill switch on the fraud classifier at 2 AM tonight, which engineer would do it, what exact steps would they follow, and when did we last verify the fallback path produces acceptable business outcomes?
- For models where the model's own decisions create selection bias in the labeled population, are we applying inverse propensity weighting to our performance metrics, and if not, how biased is our estimate of true model performance?
- How do we currently distinguish training-serving skew from genuine concept drift when a feature PSI threshold is crossed, and does the runbook document the diagnostic steps for each case?
- When a challenger model completes shadow scoring with better aggregate metrics than the champion, what segment-level and calibration checks do we require before promoting, and who has the authority to override those checks under deadline pressure?

6.8 Key Takeaway

Keeping a traditional ML model production-ready requires two parallel monitoring tracks: an online track that watches input and output distributions in real time using PSI, KS tests, the Wasserstein distance, and MMD, and an offline track that computes ground-truth performance metrics as labels arrive through a properly engineered label-arrival pipeline. Neither track alone is sufficient because online signals are fast but indirect, while offline signals are accurate but lag behind the label arrival delay. The operational stack surrounding these tracks, span-level latency instrumentation, shadow scoring, blue/green rollouts with automated step gates, and a kill-switch procedure that any on-call SRE can execute without ML engineer involvement, converts monitoring signals into safe, reversible production decisions. Slice-based and segment-level monitoring is not optional: aggregate metrics hide segment-level degradation, which is typically the earliest detectable failure mode in a production classifier; the feature store, label arrival pipeline, and inference logger must be treated as first-class engineering systems for the monitoring program to function under real production conditions.

7 LLM Evaluation and Generative AI Monitoring in Production

7.1 Opening Scenario

It is 2:47 PM on a Tuesday when Priya, a senior ML engineer on the customer-support platform team, opens a Slack thread with forty-seven messages already accumulating. Users are reporting that the support LLM is giving vague, circular answers to billing questions. CSAT scores for LLM-assisted tickets have dropped four points over ten days. Infrastructure is clean: p99 latency is nominal, token throughput is within budget, and model endpoint error rate is zero. By every signal in the existing APM dashboard, nothing appears to be wrong.

Priya pulls the inference trace logs for the past two weeks and notices that the active prompt template version changed eight days ago. A one-line addition to the system prompt, intended to make responses more concise, went out without a quality gate. She configures a sampled LLM-as-judge evaluation pipeline on a golden dataset of 200 billing-query examples with known correct answers. The rubric covers factuality, completeness, and grounding against the support knowledge base. Results return within forty minutes. Factuality scores for the current prompt are 0.61; the prior version scored 0.84. The conciseness instruction suppressed the enumeration of policy details, which turns out to be exactly what users need in billing disputes.

Priya raises a pull request to revert the prompt template, routes the rollback through the same canary deployment gate used for code changes, and flips the feature flag to cut traffic back to the prior version. CSAT scores recover over the next forty-eight hours. The post-mortem adds one mandatory step: every prompt change must pass a golden-dataset eval before reaching production. That requirement transforms a one-time fix into a durable engineering control.

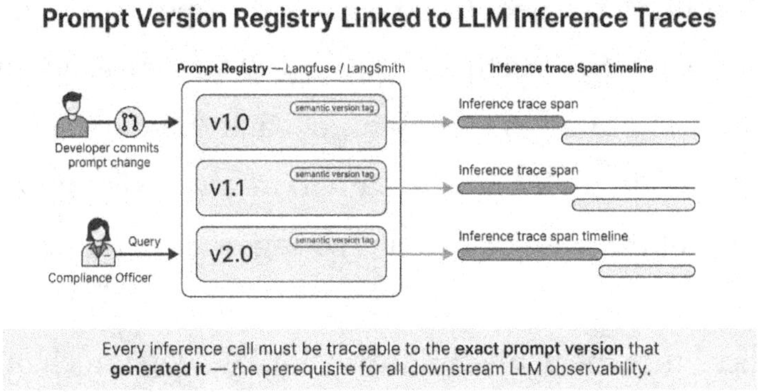

Prompt Version Registry Linked to LLM Inference Traces

Every inference call must be traceable to the **exact prompt version** that **generated it** — the prerequisite for all downstream LLM observability.

Prompt template version control pipeline: diff, canary gate, and golden-dataset eval run before production promotion

7.2 Why It Matters for Production Systems

Classical machine learning models fail in ways that infrastructure can observe directly. A gradient-boosted classifier either emits a prediction or returns an error. Its output is a scalar or probability vector. Drift detectors compare output distributions against a reference window using the Population Stability Index or a Kolmogorov-Smirnov test, and ground-truth labels close the evaluation loop automatically. Large language models break every assumption in that loop. The output of an LLM is a natural-language artifact of arbitrary length. There is no single correct answer to most prompts, no scalar to compare against a reference distribution, and no ground-truth label arriving automatically from downstream. A model can produce a response that is fluent, coherent, and formatted exactly as the prompt specifies while simultaneously being factually incorrect, regulatory noncompliant, and subtly harmful. Infrastructure telemetry surfaces none of this. Token count normal, latency nominal, HTTP status 200.

The production consequences compound over time. Output quality erodes gradually, surfacing first as anecdotal user complaints, then as aggregate degradation in satisfaction scores, then as documented business impact. By the time degradation is visible in a business metric, it has often been accumulating for weeks. The cause is almost always a change to the prompt template, a shift in the distribution of incoming queries, a change in the retrieval corpus feeding a RAG pipeline, or staleness in the knowledge base. None of these are visible without purpose-built LLM observability instrumentation.

A production LLM system is rarely a single model endpoint. It is a pipeline: a retrieval stage that fetches documents from a vector index or embedding store, a prompt-assembly stage that injects retrieved context into a template, an inference stage that calls the model API, and post-processing stages that filter or format the output. Each stage can fail independently. Retrieval recall can drop because the embedding store is stale. Context injection can truncate critical information when the context window limit is reached. The model can hallucinate details absent from the retrieved context.

Eval drift adds a third failure mode. A golden dataset built at launch reflects the query distribution and quality expectations of the launch cohort. Six months later, users may be asking qualitatively different questions, the regulatory environment may have changed, and the application's intended scope may have expanded. If the eval harness is not updated, it continues reporting high scores while the application silently diverges from current requirements. The engineer's task is to close these gaps with a layered observability architecture: an eval layer that scores output quality against defined rubrics; an instrumentation layer that tracks every quality dimension across the full pipeline; and a guardrail layer that enforces input and output constraints in real time and records violations as telemetry events. LLM evaluation at scale incurs real inference costs, so the architecture must include an explicit eval-cost budget and a stratified sampling strategy that concentrates evaluation effort where the risk of undetected quality failure is highest.

7.3 Online and Offline Eval Architecture

A mature LLM observability program runs both offline and online evaluation. Offline eval answers whether a new prompt version, model version, or retrieval configuration is better or worse than the current baseline before production promotion—online eval answers whether the production system is performing within acceptable quality bounds at the moment. Online eval failures trigger offline investigations, and offline eval gates are enforced before online exposure.

Golden Datasets and Eval Harness Design

A golden dataset is a curated collection of input-output pairs representing the application's intended behavior at a point in time. Treat its construction as a prerequisite to launch, not a post-launch activity. A minimum viable dataset of 50 to 100 examples enables regression detection across prompt versions, which is the primary value at launch. The dataset is a versioned code artifact requiring quarterly updates from production traffic samples to prevent eval drift. The eval harness runs evaluation jobs, manages the golden dataset, stores results in a queryable database, and generates comparison reports between versions. Tools like Langfuse, Braintrust, or LangSmith provide managed eval harness infrastructure with built-in support for LLM-as-judge evaluators, custom metric functions, and version-to-version comparison dashboards. The harness must be wired into the CI/CD pipeline so every prompt template change automatically triggers an eval run. A configurable quality gate, for example, requiring that the factuality score not drop more than 0.05 from baseline, blocks regressions from reaching production.

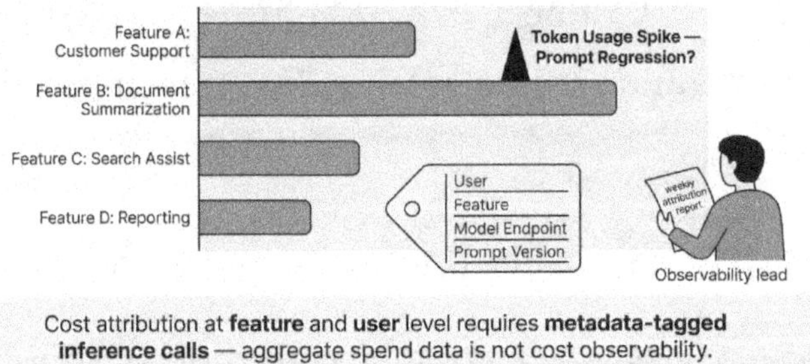

Offline eval harness: golden dataset versioning, eval job runner, metrics store, and CI/CD quality gate

LLM-as-Judge, Rubric Scoring, and Statistical Aggregation

LLM-as-judge is the technique of using a separate language model to score another model's outputs against a structured rubric. The judge receives the original query, the system prompt, any retrieved context, and the produced response, and returns a scoring object with a numeric score for each rubric dimension and a natural-language rationale for each score. Effective rubrics define each dimension operationally: for factuality in a RAG application, a working definition might read, "The response makes no claims that contradict information in the retrieved documents, and all material claims are supported by at least one retrieved chunk." Rubric dimensions for most production LLM applications span at a minimum factuality, completeness, instruction-following, and refusal correctness. Each dimension gets its own time-series score, never collapsed into a single composite. Judge models exhibit scoring variance; track the standard deviation of judge scores over time as a separate metric, and spot-check the judge weekly against expert human annotation to catch scoring drift before it corrupts the eval program.

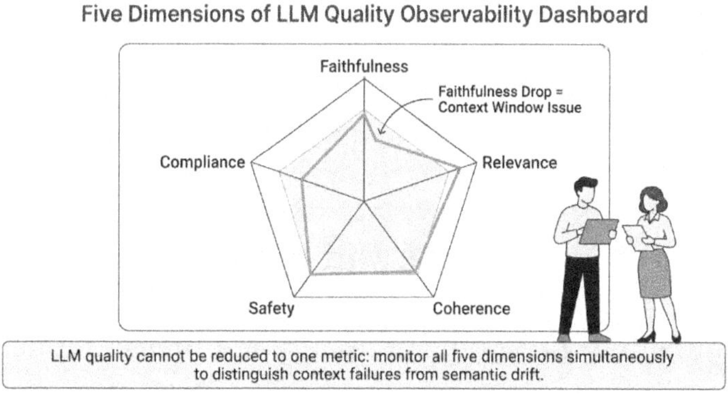

Five Dimensions of LLM Quality Observability Dashboard

LLM-as-judge scoring pipeline: input bundle construction, judge call, structured score and rationale output, human-calibration loop

Sampling Strategies and Eval-Cost Budgeting

Stratified sampling is more effective than uniform random sampling for online evaluation. Draw a baseline random sample of one to five percent of total traffic, and additionally oversample specific risk strata: outputs with low automated proxy scores, outputs from sessions where users triggered dissatisfaction signals, outputs generated during active semantic drift or retrieval-quality alerts, and all outputs from new prompt template versions during their canary window. Eval-cost budgeting should be a first-class engineering constraint allocated as a line item in the total inference cost budget, with a reasonable starting point of five to ten percent of total inference cost. Cost can be reduced by using a smaller, cheaper model as the primary judge for high-volume dimensions, such as instruction-following, while reserving the more capable judge for high-risk dimensions, such as factuality and refusal correctness. Tools like Braintrust and Arize Phoenix support configuring different evaluator models per rubric dimension.

7.4 Quality Dimensions and Instrumentation

Instrumenting LLM output quality requires a defined set of measurable dimensions that together characterize production health from the user's perspective.

Factuality, Faithfulness, Hallucination Detection, and Retrieval Quality

Factuality measures whether the model's output is true relative to an external ground truth. Faithfulness, also called groundedness, measures whether the output stays within the scope of information in the context window and does not introduce parametric-knowledge claims that contradict or extend beyond the retrieved context. Hallucination detection identifies outputs containing plausible-sounding but factually unsupported claims. For RAG applications, it can be implemented by comparing each factual claim in the response against retrieved documents using an entailment classifier or an LLM-as-judge evaluator. The hallucination rate, the fraction of evaluated responses containing at least one unsupported claim, is tracked as a time-series metric with a configurable alert threshold. Citation grounding links each response claim to the specific retrieved chunk that supports it; the citation map, stored in the inference trace, pinpoints whether a hallucination failure originated in the retrieval stage or the generation stage.

For RAG applications, retrieval quality is a first-class observability concern. The core metrics are recall@k, normalized Discounted Cumulative Gain (nDCG), and Mean Reciprocal Rank (MRR). Recall@k measures the fraction of relevant documents in the top-k retrieved results. nDCG is rank-weighted, giving greater credit to relevant documents that appear earlier in the list. MRR measures the average reciprocal rank of the first relevant document. A drop in recall@k or nDCG below the deployment baseline signals embedding store staleness or vector index corruption. Embedding drift on the retrieval corpus is a related failure mode: as documents are added, removed, or modified, the distribution of document embeddings shifts. The standard monitoring approach computes a reference embedding distribution at deployment time, then runs daily comparisons using the Mahalanobis distance or the maximum mean discrepancy, alerting when the distances exceed a configured threshold.

LLM-as-a-Judge Evaluation Pipeline Architecture

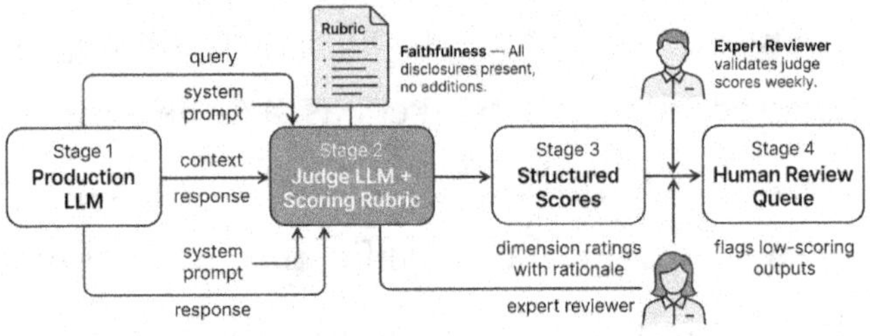

The judge model must be governed like a **production system:** its own version, prompt, and weekly calibration **against human** expert review.

Faithfulness instrumentation for RAG: claim extraction, retrieved-chunk alignment, citation map construction, and hallucination rate aggregation

Refusal Correctness, Toxicity, Bias, and PII Leakage

Refusal correctness measures whether the model refuses requests it should refuse and answers requests it should answer. Measuring it requires a two-sided evaluation dataset: requests that should be answered (refusal is a false positive) and requests that should be refused (answering is a false negative). The jailbreak rate is the key safety metric. Toxicity event rates tracked by prompt version and user segment expose two patterns: an uptick correlated with a prompt version change signals that the change weakened a safety constraint, and an uptick correlated with a specific user segment may indicate coordinated adversarial behavior.

PII leakage monitoring instruments the output stream for personally identifiable information. Structured PII, such as Social Security numbers and email addresses, can be detected using regular expression patterns. Unstructured PII requires named entity recognition (NER) models running as a post-processing filter. The PII event rate, segmented by PII type and prompt version, is a required time-series metric. Organizations operating under GDPR, HIPAA, or CCPA requirements cannot rely on the base model to self-censor; a production guardrail filter scanning all outputs is a compliance requirement, not an optional enhancement.

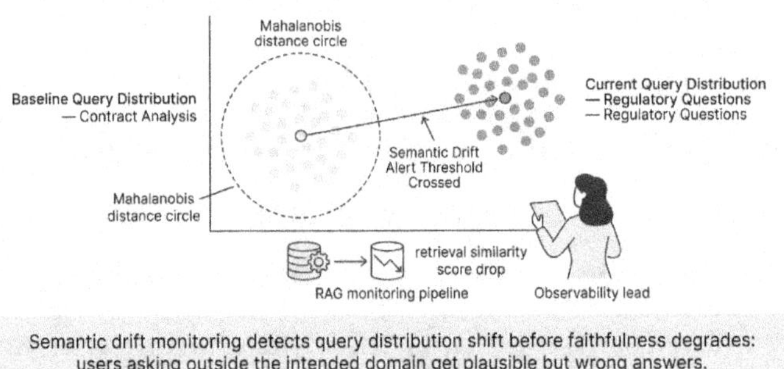

Semantic drift monitoring detects query distribution shift before faithfulness degrades: users asking outside the intended domain get plausible but wrong answers.

Quality dimension instrumentation panel: factuality, refusal-correctness, toxicity rate, PII event rate, and hallucination rate per prompt version

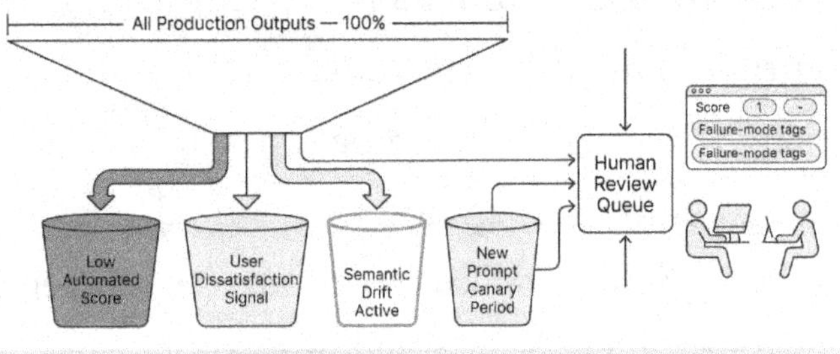

Stratified sampling concentrates human review effort where undetected quality failure risk is highest, maximizing diagnostic value per review hour.

Retrieval quality monitoring: recall@k and nDCG trend lines against deployment baseline, with embedding-drift alert overlaid

7.5 Production Guardrails and Continuous Monitoring

Guardrails are the real-time enforcement layer that sits between the LLM inference call and the user response. A guardrail filter runs inline on every request and response, blocking or transforming content that violates configured policies. Guardrails catch known bad patterns at the boundary; the eval layer detects subtle quality degradation in the interior. Both are required.

Input and Output Guardrail Architecture

The input guardrail layer evaluates user-supplied text before prompt assembly, performing prompt-injection detection, input toxicity filtering, and PII redaction. Prompt-injection detection uses classifiers trained on known injection patterns, semantic similarity against a library of known attack templates, and output-side monitoring for responses indicating manipulation. No single classifier achieves perfect recall on novel injection patterns, so the architecture must be defense-in-depth: the input guardrail catches high-confidence injections. It logs them as telemetry events, while the output guardrail catches injection-success signatures such as the model outputting its own system prompt or generating content inconsistent with the configured persona. All injection events are stored in the audit log, including the full input text, for forensic analysis. Output guardrails enforce PII suppression, toxicity blocking, and response-format validation. Each filter's decision and latency are recorded as a span in the trace, making the guardrail layer debuggable.

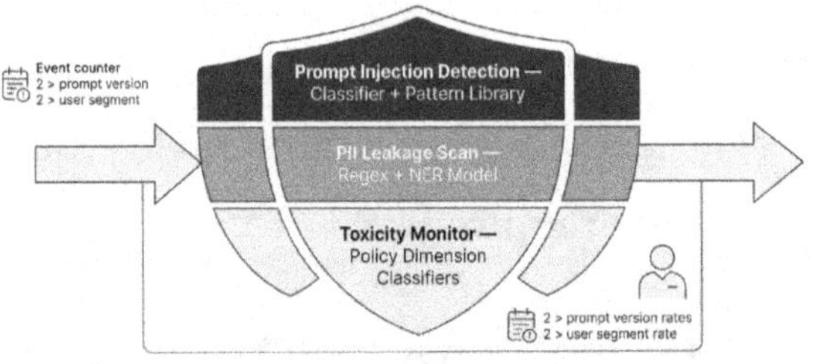

Inline guardrail architecture: input-stage injection detection, output-stage PII filter and toxicity classifier, with span instrumentation per filter layer

Response Distribution Drift, Prompt Versioning, and Semantic Cache Monitoring

Response distribution monitoring tracks statistical properties of the model's output stream over time, independent of semantic quality scores. Useful signals include the response-length distribution, the response-format compliance rate, and the refusal rate. A sudden shift toward shorter responses may indicate a prompt change that suppresses elaboration; a rising refusal rate without a corresponding change in query distribution often signals a prompt regression that over-activates safety instructions. Prompt template versioning is the foundational control that makes response distribution monitoring actionable: without it, a drift event in response length has no attributed cause. Every inference call must include the active prompt template version as a metadata tag in the trace and the metrics event. When a new version is deployed, the system computes a prompt template diff against the prior version and surfaces it as a deployment annotation in the monitoring dashboard. Tools like Langfuse, LangSmith, and Braintrust support prompt version tagging and diff views. A semantic cache stores embeddings of previously answered queries alongside their responses, returning cached responses when incoming queries fall within a configured similarity threshold. The cache hit rate and similarity threshold are observable metrics; a sudden drop in hit rate is an early signal of semantic drift before it appears in quality metrics.

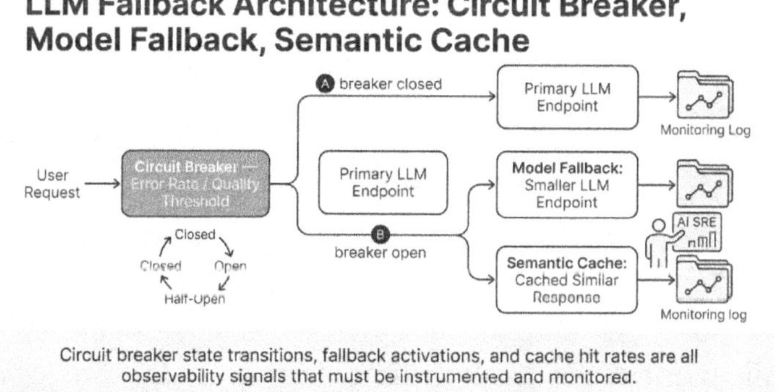

Circuit breaker state transitions, fallback activations, and cache hit rates are all observability signals that must be instrumented and monitored.

Response distribution drift monitoring: length histogram by prompt version, refusal-rate time series, semantic cache hit rate, and diff annotation on deployment events

Continuous Monitoring Cadence and the LLM Health Review

The monitoring cadence should operate on three time scales. Within minutes, the inline guardrail layer logs every injection event, PII block, and toxicity flag as structured telemetry with alert thresholds for sudden spikes. Within hours, a lightweight online eval job processes the day's sampled traffic through the LLM-as-judge pipeline, updating quality dimension time series; a regression alert fires if any dimension drops more than two standard deviations below the rolling seven-day baseline. Within days, the offline eval harness runs a full evaluation on the golden dataset against any new prompt template or retrieval configuration deployed in the preceding window.

The weekly LLM health review is the human governance loop that closes the observability cycle. The ML platform engineer, the model owner, and an SRE with on-call context review quality dimension trends, retrieval quality metrics, guardrail event rates, embedding drift signals, and prompt template changes deployed. The review produces a written health record: current status per dimension, interventions decided, and owners with timelines. This record is the audit trail demonstrating active management of LLM application risk; for applications operating under AI governance frameworks, it is the primary artifact regulators and internal audit functions will examine. Patterns surfaced here feed back directly: quality degradation patterns drive prompt engineering work, retrieval quality gaps drive knowledge base expansion, and safety event patterns drive guardrail enhancements.

Weekly LLM Health Review: Governance Cycle and Health Record

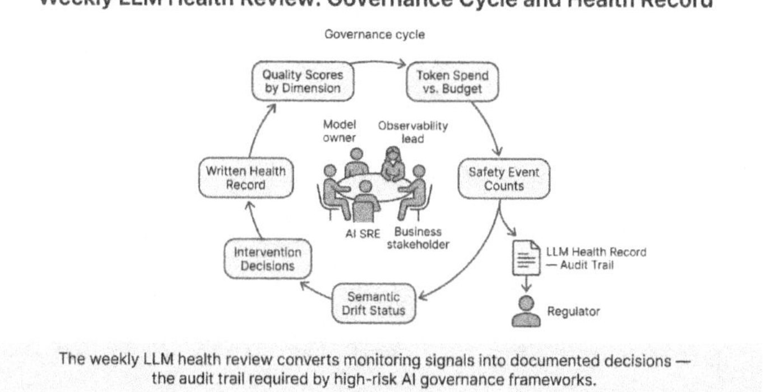

The weekly LLM health review converts monitoring signals into documented decisions — the audit trail required by high-risk AI governance frameworks.

Continuous monitoring cadence: real-time guardrail telemetry, hourly eval aggregation, daily retrieval quality report, and weekly health review cycle

7.6 Technical Team Checklist: LLM Eval and Monitoring

- **Action:** Establish a prompt template registry and treat every change as a versioned deployment requiring a pull request, a diff annotation in the monitoring system, and a golden-dataset eval run before promotion. **Owner:** ML Platform Eng. **Done when:** No prompt template reaches production without a version tag, a stored diff, and a passing eval gate in CI/CD.

- **Action:** Instrument every inference call with structured metadata tags for user identifier, feature name, model endpoint, and active prompt template version, emitting these as span attributes on the distributed trace. **Owner:** ML Engineer. **Done when:** The metrics backend can filter any quality or cost time series by prompt version, feature, and model endpoint with no manual joins.
- **Action:** Build and maintain a golden dataset of at least one hundred input-output pairs per application domain, versioning it as a code artifact and updating it quarterly from production traffic samples with domain expert annotation. **Owner:** ML Engineer. **Done when:** The eval harness runs golden-dataset evals on every prompt change and stores version-to-version comparison reports in the metrics store.
- **Action:** Deploy an LLM-as-judge eval pipeline with operationally defined rubrics covering factuality, faithfulness, completeness, refusal correctness, and instruction-following, with separate scores per dimension stored as time-series metrics. **Owner:** ML Platform Eng. **Done when:** Quality dimension scores are queryable in the monitoring dashboard alongside latency and error rate, with alert thresholds configured per dimension.
- **Action:** Implement retrieval quality monitoring for all RAG applications, computing recall@k, nDCG, and MRR

daily against a retrieval evaluation dataset and alerting when any metric drops more than ten percent below the deployment baseline. **Owner:** ML Engineer. **Done when:** Retrieval quality metrics are trending in the monitoring dashboard, and an on-call alert is configured for the retrieval quality SLO.

- **Action:** Wire inline input and output guardrail filters for prompt-injection detection, PII suppression, and toxicity classification, instrumenting each filter as a span in the distributed trace with logged disposition and latency. **Owner:** ML Platform Eng. **Done when:** Injection event rate, PII block rate, and toxicity flag rate are queryable as time-series metrics with per-prompt-version segmentation.
- **Action:** Implement embedding drift monitoring on the retrieval corpus, computing distributional distance against the deployment-time reference embedding distribution daily and alerting when distance exceeds two standard deviations from the baseline. **Owner:** ML Platform Eng. **Done when:** Embedding drift distance is a tracked metric in the monitoring dashboard, and the re-indexing runbook references the drift alert as its primary trigger.
- **Action:** Define and enforce an eval-cost budget as a fraction of total inference cost, configure the sampling policy to oversample high-risk strata (low proxy scores,

user dissatisfaction signals, canary windows), and track eval cost monthly. **Owner:** ML Engineer. **Done when:** The eval budget is documented, the sampling policy is implemented in the eval harness, and the monthly eval cost is visible in the FinOps dashboard.

- **Action:** Conduct a weekly LLM health review covering quality dimension trends, retrieval quality, guardrail event rates, embedding drift status, and prompt version changes, producing a written health record per session. **Owner:** ML Platform Eng. **Done when:** Health records exist for every weekly review session and are retrievable on demand for governance audit.

7.7 Team Conversation Starters

- If a prompt template changed last week and quality metrics dropped two days later, what is the minimum instrumentation needed to correlate those events unambiguously, and does the current system provide it?
- For the highest-volume user query clusters, what is the current recall@k for the retrieval stage, and is that metric trended anywhere the on-call engineer can reach during an incident?
- When the LLM-as-judge scores a response as low-faithfulness, does the team have a citation map identifying whether the failure originated in retrieval or

in generation, and how long would it take to answer that question for an arbitrary production incident?

- If a coordinated prompt-injection campaign were launched against the customer-facing application tonight, what signals would fire in the on-call queue, within what time-to-detection, and has that detection path been tested end-to-end?
- The golden dataset was built at launch. Have the query distribution, business requirements, and regulatory context changed enough that the eval scores it produces may be measuring something different from what the team thinks they are measuring?
- If the retrieval corpus embedding model is updated by its provider without notice, what would the observability system detect first, and is there a runbook for that scenario?

7.8 Key Takeaway

LLM observability is a purpose-built engineering discipline, not a retrofit of classical ML monitoring. The failure modes are qualitative, the ground truth is contested, and the causal chain from a prompt template change to a user-visible quality regression can span days and multiple pipeline stages. The architecture that makes this tractable combines three interconnected layers: a versioned eval harness running golden-dataset and LLM-as-judge evals on a defined sampling budget; a quality instrumentation layer tracking factuality, faithfulness, retrieval quality, refusal correctness, and PII and toxicity event rates as first-class time-series metrics segmented by prompt version; and an inline guardrail layer recording injection, PII, and toxicity events as span telemetry. Closing the loop requires a weekly human review that converts these signals into prioritized engineering actions and a written audit record. Teams that build all three layers and connect them through a consistent prompt versioning discipline detect LLM quality regressions in hours rather than weeks.

8 Agent Observability: Tracing and Governing Agentic AI Systems

8.1 Opening Scenario

Priya, the ML platform engineer on call for the research automation platform, is watching the cost dashboard on a Thursday afternoon when a line bends upward at a slope that should not exist in a single-agent workflow. The autonomous research agent, spun up ninety minutes ago to pull competitor pricing data, has logged 1,847 tool calls. The running cost counter reads $4,217.

She opens the raw log stream and immediately sees the pattern: the same API call, over and over, with a slightly mutated query string each time. The agent is retrying a fetch against a structured data endpoint that keeps returning HTTP 200 with a body that is not valid JSON. The upstream service returns a truncated HTML error page when its backend is overloaded. The agent has no schema validation on the tool output. It sees the 200 status, interprets the response as success, cannot parse the result, and retries with a slightly rephrased query. It has been doing this for eighty-eight minutes.

The cost-per-task sentinel fires when the workflow crosses the $4,000 hard kill threshold. Total cost: $4,217.43. Useful output: zero.

Priya opens the trace store. The root span is there. Its 1,847 child spans show HTTP 200 responses with a uniform 340-millisecond duration. None carries a validation-failure attribute. None captures the raw response body. The trace reports that 1,847 tool calls were completed. It does not reveal that every result was malformed.

The fix takes eleven minutes: add output schema validation to the tool-call span instrumentation, surface an `agent.tool.output.valid` boolean on every tool-call span, and add a loop-detection guard that trips after ten consecutive validation failures on the same tool. The observability gap that cost over $4,000 was due to one missing span attribute and one missing circuit breaker.

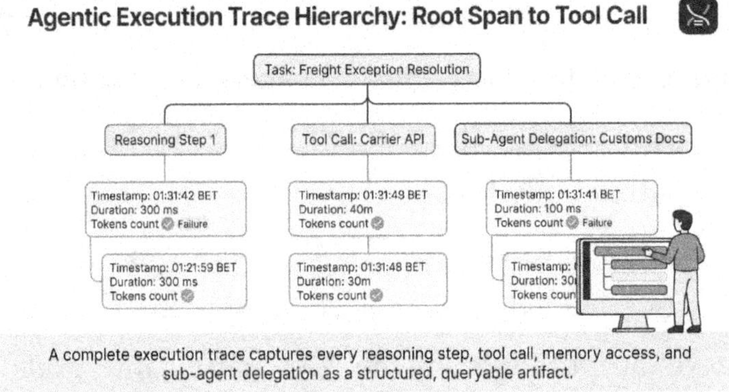

A complete execution trace captures every reasoning step, tool call, memory access, and sub-agent delegation as a structured, queryable artifact.

Cost-sentinel trace: 1,847 tool-call spans with uniform HTTP 200 status masking malformed output, annotated at the kill-switch trigger point

8.2 Why It Matters for Production Systems

Traditional software monitoring assumes a finite, enumerable state space. The error taxonomy is tractable, and you can build a complete monitoring stack around a known behavioral contract. Agentic AI systems invalidate this assumption at the architectural level. An agent is not a function that maps input to output. It is a process that reasons, decides, and acts across a sequence of steps, where each step can branch into any number of subsequent steps, the set of possible actions is partially open-ended, and the relationship between any individual step and the outcome is not deterministic. The monitoring primitives that work for web services and even for LLM prompt-response pairs do not transfer to agent loops without significant extension.

Agents fail in ways invisible to output-only monitoring. The error almost always lives in the middle of the execution trace: a tool that returned an unexpected structure, a reasoning step that made a faulty inference. This sub-agent interpreted its delegated task differently from the orchestrator's intention. If the observability stack captures only task start, task end, and final output, those intermediate failure modes are unrecoverable. Metrics alone are insufficient because agentic failure modes do not respect aggregation boundaries. A multi-agent workflow running 200 sessions per day might show a 94 percent aggregate success rate, while a specific sub-agent has been silently hallucinating citations on a particular query class for weeks. A per-agent, per-task-type, per-tool success rate wired to trace-level replay for failing sessions surfaces this in the first incident review.

The blast radius problem is unique to agentic systems. When a traditional model makes a bad prediction, downstream harm is proportional to how that prediction was used. When an agent executes a bad action sequence, harm can propagate through multiple downstream systems before any human is aware. An agent writing to a shared data store with incorrect field values corrupts every downstream pipeline reading from that store. The cost dimension is equally non-trivial: a classification model's cost per inference is essentially constant, while an agent's cost per task can vary by orders of magnitude depending on tool call volume and whether it enters loops. An agent entering a behavioral loop generates no runtime exception or error-rate alert; the only signal is the pattern of behavior over time: the same tool called repeatedly with marginally different parameters, cost accumulating without progress, and task duration far exceeding historical p99 for that workflow type.

8.3 Trace Architecture for Agentic Systems

The foundational instrumentation primitive for any agentic system is the hierarchical trace with typed spans. An OpenTelemetry trace for an agent task has a root span that represents the entire task execution, with child spans for each logical step in the agent loop. The standard step taxonomy follows the plan/act/observe cycle: a plan span records the agent's reasoning, including the prompt sent and completion received; an act span records the tool call with input parameter capture; and an observe span records the agent's processing of the tool result, including schema validation and the interpretation the agent carries forward.

Each span in a ReAct trace should carry a standard attribute set. The OpenTelemetry GenAI semantic conventions provide a baseline: `gen_ai.system`, `gen_ai.request.model`, `gen_ai.usage.input_tokens`, `gen_ai.usage.output_tokens`, `gen_ai.operation.name`. For tool-call spans, the platform team extends with agent-specific attributes: `agent.tool.name`, `agent.tool.input` (serialized JSON), `agent.tool.output` (serialized JSON), `agent.tool.output.valid` (boolean from schema validation), `agent.tool.latency_ms`, `agent.loop.depth`, and `agent.task.id` (stable identifier linking all spans in a session). Span attributes for tool args are what distinguish a debugging artifact from a log entry: they record what the tool was called with, what it returned, whether the return value was valid, and how deep in the agent loop the call fired. These attributes drive root cause analysis in minutes rather than hours.

Agent Decision Graph Visualization for Management Review

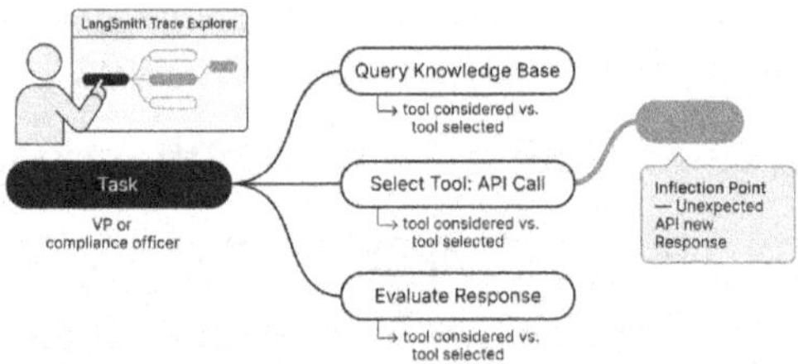

Decision graph visualization must be accessible to non-engineers — a graph readable only by builders is a debugging tool, not a governance artifact.

OpenTelemetry span hierarchy for a ReAct loop iteration: root task span, child plan span with token counts, child tool-call span with input/output JSON and validation flag, child observe span with extracted result

Tool-Call Instrumentation and Span Attributes for Tool Args

Every tool in the agent's toolset should be wrapped in a span-emitting decorator at the tool-registry layer that fires before and after the tool executes, capturing input arguments, output content, HTTP status, latency, and the schema validation result. This instrumentation lives at the registry level so that new tools are automatically covered. A secrets-scrubbing middleware in the export pipeline handles credential and PII redaction before spans reach the trace store.

The `agent.tool.output.valid` boolean is the single most important agent-specific span attribute. Schema validation of tool output should be mandatory for every tool that returns structured data. On validation failure, the span attribute should record a short error string describing the violated schema constraint, converting "why did the agent do X" into "the tool returned a response missing the required `results` field." Capturing that validation state in the span is the difference between a four-hour debugging session and an eleven-minute fix.

Span Linking Across Multi-Agent Handoffs

When an orchestrator delegates a subtask to a worker agent, the worker runs in its own process and generates its own root span. Without explicit span linking, the orchestrator's trace shows a delegation event and a result with no connection to the worker's intermediate reasoning. An incident originating in the worker is invisible at the orchestrator level without a shared parent span context.

The standard solution is W3C Trace Context propagation via the `traceparent` header. When the orchestrator delegates, it passes its current span context. The worker creates its root span with that context as the parent, establishing a parent-child relationship across process boundaries. In the trace store, the unified view shows a single span tree: orchestrator root, planning and delegation spans, worker spans as children of the delegation span. Frameworks without native W3C support require a shim at each handoff boundary. Tools like AgentOps and LangSmith support span linking natively; mixed-framework deployments require explicit propagation code. The platform team should verify end-to-end trace correlation in staging before any multi-agent system goes to production.

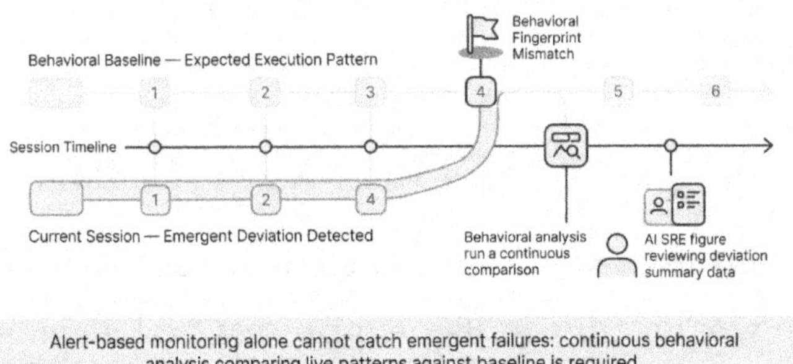

Multi-agent span linking: orchestrator root with child delegation spans, W3C traceparent propagation to worker agents, unified trace tree showing cross-process parent-child relationships.

Deterministic Replay from Trace for Debugging

Deterministic replay means re-running the agent against exactly the inputs and tool outputs it saw during the original execution, without making live tool calls or consuming new tokens. The trace store contains, for each tool-call span, the serialized input arguments and the serialized output. A replay harness stubs tool implementations with functions that return the recorded outputs in sequence and re-runs the agent's reasoning logic against those fixed outputs. The engineer modifies recorded tool outputs one at a time to isolate the exact response that triggered the failure. An agent that failed at 2 AM due to a tool returning malformed JSON can be debugged at 10 AM using only the trace. Teams using LangSmith or Langfuse can use their built-in trace replay features; teams with custom infrastructure should build this capability during initial platform buildout.

8.4 Behavior Monitoring and Safety Controls

Instrumentation produces traces. Safety requires acting on what those traces reveal. The behavioral monitoring layer is the set of automated checks running against the live trace stream, detecting pathological patterns and triggering interventions before the blast radius grows. The cost sentinel in the opening scenario is the minimum viable safety system. Production systems need substantially more.

Loop detection is the most fundamental behavioral safety control. An agent entering an agent loop is not in an error state from the runtime's perspective but exhibits a behavioral pattern that, left unchecked, consumes arbitrary resources. Loop detection should operate at multiple granularities: at the span-attribute level, tracking whether the same tool is called with substantially similar inputs within a rolling window; at the trace-depth level, counting total tool-call spans within a single task and tripping at a configured maximum; and at the time level, alerting when a task has been running longer than the p99 of historical duration.

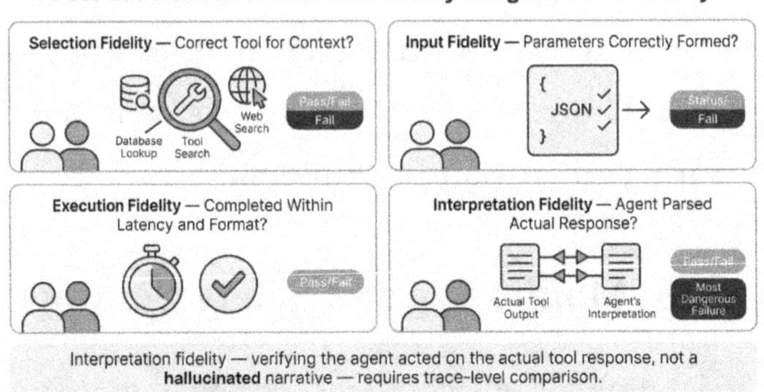

Loop detection signal stack: span-level tool-call similarity monitor, trace-depth counter, elapsed-time alert, and cost-per-task sentinel forming a four-layer defense against runaway agent loops

Recursion Depth Caps, Rate Limits, and Output Validation

Recursion depth caps enforce a hard ceiling on agent nesting depth. The `agent.loop.depth` span attribute enables enforcement: the agent framework checks the attribute at each delegation step and refuses to create a new sub-agent if the current depth exceeds the configured cap. A reasonable default is four levels of nesting; deeper hierarchies should document the justification for the deeper nesting. Tool-call rate limits restrict the rate at which any agent can invoke a particular tool within a time window, protecting downstream APIs and bounding the maximum cost any single agent can accumulate in a loop. They should be implemented at the tool-registry layer so that any call that would exceed the limit is rejected before reaching the external API.

Tool-output validation is non-negotiable for any tool that returns structured data. Every tool in the authorized toolset should have a registered JSON Schema or Pydantic model, with validation running immediately after the tool returns and before the output is passed to the reasoning layer. On validation failure, the platform surfaces a structured error response to the agent so it can handle the failure gracefully. Passing malformed output to the LLM and letting it improvise an interpretation is the default behavior in most agent frameworks without explicit instrumentation, and it is the failure mode this chapter opened with.

Scope Drift Detection, Action Allowlisting, and Sandboxing

Scope drift is the behavioral pattern in which an agent begins taking actions outside the boundaries of its defined task authority. The action allowlist is the primary enforcement mechanism: the platform team defines an explicit list of tools the agent is authorized to call, enforced at the tool-registry layer, so any call to an unlisted tool is blocked and generates a scope-violation span event that triggers an on-call alert. A pattern of scope-violation events warrants a review of the system prompt and the task distribution. Semantic drift detection supplements allowlist enforcement by applying an LLM-as-a-judge evaluator to the session-level trace, routing flagged sessions to a human review queue rather than terminating them automatically.

Sandboxed tool environments isolate the external effects of agent tool calls during testing by routing calls at the registry layer to mock implementations that return realistic yet synthetic responses, with no risk of production-side effects. Kill-switch design requires a graceful shutdown sequence: the agent framework receives the kill signal, completes in-flight tool calls that cannot be safely interrupted, records a terminal span event with the kill reason, and runs registered cleanup handlers.

Session-Level Agent Evaluation: Coherence and Goal Alignment

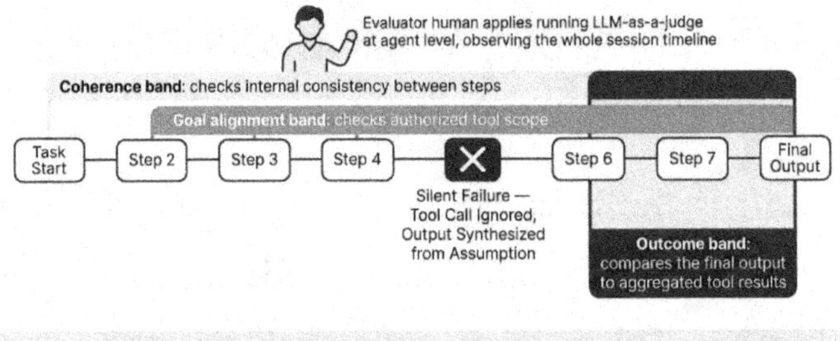

Session-level evaluation distinguishes an agent that recovered from a tool failure from one that silently fabricated output — only whole-session analysis reveals the difference.

Scope enforcement stack: a tool allowlist check at the registry layer blocking unauthorized calls, a semantic drift evaluator applied to session-level traces, and a human review queue for semantically anomalous sessions

8.5 Cost and Outcome Telemetry

An agent that runs correctly and produces good output is still a production liability if its cost and resource consumption are unmonitored. Agent cost is highly variable, can spike orders of magnitude above baseline without any visible error condition, and the aggregated cost of hundreds of concurrent agents can make a platform economically unviable overnight.

Cost-per-task is the fundamental unit-cost metric. It should be computed and recorded as a span attribute on the root span of every task execution, summing the cost contributions of all LLM calls, billable tool calls, and retrieval operations. The p50, p95, and p99 cost-per-task percentiles for each workflow type should be visible on the same dashboard as latency percentiles, with alerts triggered when cost-per-task exceeds the p99 by a configurable multiple.

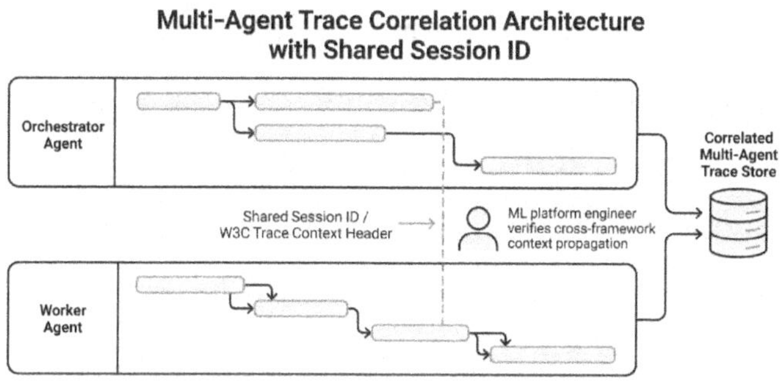

Without shared trace context propagation, an incident originating in a worker agent is invisible at the orchestrator level.

Cost-per-task telemetry: real-time cost accumulation per task overlaid on historical p50/p95/p99 distribution, with cost-sentinel threshold annotated at the kill-switch trigger point

Tokens-per-Task, Tool-Call Success Rate, and Retrieval Relevance

Tokens-per-task is the upstream leading indicator for cost-per-task. When token consumption per task increases before cost does, it signals the agent is doing more reasoning work, possibly because it is receiving noisier inputs or because its context window is filling with irrelevant retrieved content. Decomposing tokens-per-task into input and output tokens for each reasoning step provides an early warning before costs escalate. Tool-call success rate measures the fraction of tool calls that complete with a response that passed schema validation within the expected latency range. An HTTP 200 with a malformed body is a tool-call failure by this definition. A rate below the threshold should trigger an alert and, optionally, a circuit breaker that turns off the tool. Retrieval relevance for agents using retrieval-augmented generation should be measured using a lightweight scorer such as a cross-encoder or embedding cosine similarity, captured as a span attribute on the retrieval span, and aggregated into a per-retriever metric monitored for drift.

Eval-as-Spec and Regression Suites for Agent Workflows

Eval-as-spec defines the agent's desired behavior as a formal evaluation suite and runs it continuously against production output. For each workflow type, the platform team specifies a set of golden trajectories representing correct behavior, including the sequence of tool calls, the intermediate reasoning steps, and the final output. Regression evals for agent workflows are the behavioral equivalent of unit tests for traditional software. A regression eval might assert: the agent should call the search tool before the synthesis tool; the agent should not call the external API tool unless the search tool returns fewer than three relevant results; the final output should cite at least two sources from the retrieved documents. These assertions are evaluated automatically against production traces, with violations triggering alerts and routing sessions to human review.

Human-in-the-Loop Oversight Checkpoints for High-Stakes Agent Actions

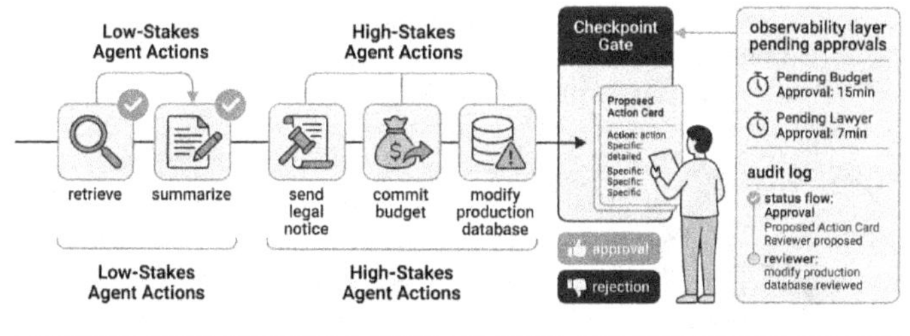

Every potentially irreversible action must pause for explicit human approval: high-stakes agent actions require an observable, audited checkpoint.

Eval-as-spec pipeline: golden trajectory library feeds a trace comparison engine evaluating production traces against behavioral assertions, with regression failures routed to the incident queue and evaluation dashboard

The success-rate-per-task metric is the aggregate outcome telemetry for the agent platform. It should be tracked separately for each workflow type, not aggregated across all tasks. A 94 percent overall success rate that hides a 30 percent failure rate on a specific task type is not a healthy signal. The platform team defines success criteria at deployment time, encodes those criteria in the eval-as-spec suite, and reports success-rate-per-task as a first-class SLO with an error budget tracked in real time. When the budget is exhausted, a freeze on non-critical changes goes into effect until the success rate recovers. The telemetry stack should consolidate cost-per-task, tokens-per-task, tool-call success rate, retrieval relevance, and task success rate into a single agent health dashboard. Tools like AgentOps, Langfuse, and Arize AI provide out-of-the-box dashboards for many of these metrics for common agent frameworks.

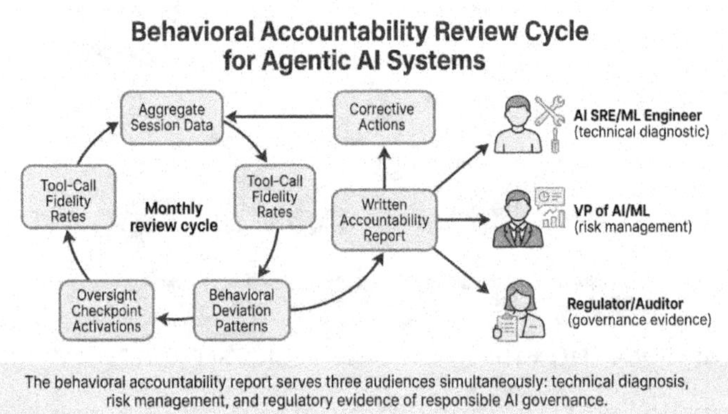

The behavioral accountability report serves three audiences simultaneously: technical diagnosis, risk management, and regulatory evidence of responsible AI governance.

Agent health dashboard: cost-per-task, tokens-per-task, tool-call success rate by tool, retrieval relevance score, and task success rate per workflow type, all sourced from the same trace attribute stream

8.6 Technical Team Checklist: Agent Tracing and Governance

- **Action:** Instrument every tool in the agent's toolset with a span-emitting decorator at the tool-registry layer, capturing `agent.tool.name`, `agent.tool.input`, `agent.tool.output`, `agent.tool.output.valid`, `agent.tool.latency_ms`, and `agent.loop.depth`, with a secrets-scrubbing middleware in the span export pipeline. **Owner:** ML Platform Eng. **Done when:** A sample of 100 production traces shows 100% attribute completeness across all tool-call spans.
- **Action:** Deploy W3C Trace Context propagation across all agent boundaries, embedding the `traceparent` header in orchestrator-to-worker delegations and verifying unified span linking in the trace store. **Owner:** ML Platform Eng. **Done when:** A multi-agent session trace in staging shows a single unified span tree with parent-child relationships across all agent process boundaries.
- **Action:** Register a JSON Schema or Pydantic model for every structured-output tool, wire schema validation into the instrumentation layer, and trigger a circuit-breaker

alert after ten consecutive `agent.tool.output.valid=false` events on the same tool within a single task. **Owner:** ML Engineer. **Done when:** A test run with a malformed tool response triggers the validation failure attribute and circuit-breaker alert within one tool-call span.

- **Action:** Implement a four-layer loop detection stack covering span-level tool-call similarity, trace-depth counting, elapsed-time alerting, and cost-per-task sentinel, with independently configured thresholds per workflow type and a graceful kill-switch that runs cleanup handlers and records the trigger reason as a terminal span event. **Owner:** ML Platform Eng. **Done when:** A synthetic infinite-loop test agent is killed cleanly at each threshold, and the kill-event span appears in the trace with the correct trigger reason.
- **Action:** Define and enforce an action allowlist for every deployed agent at the tool-registry layer, blocking any call to an unlisted tool and generating a scope-violation span event that routes to the on-call alert channel. **Owner:** Governance Eng. **Done when:** A test agent attempting an unauthorized tool call is blocked, a scope-violation event appears in the trace, and the on-call alert fires within sixty seconds.
- **Action:** Build a deterministic replay harness that loads a stored trace, stubs tool implementations with recorded outputs, and re-runs agent reasoning against fixed

inputs, shipped as a standard developer CLI command. **Owner:** ML Platform Eng. **Done when:** Any engineer can replay any stored trace from the command line in under five minutes without network access to live external systems.

- **Action:** Define golden trajectories for each production workflow type, encode behavioral assertions as an automated regression eval suite, and run the suite nightly against a sample of production traces. **Owner:** ML Engineer. **Done when:** The regression eval suite runs in CI with pass/fail status per workflow type, and a regression failure blocks agent configuration changes.
- **Action:** Wire cost-per-task, tokens-per-task, tool-call success rate per tool, retrieval relevance per retriever, and task success rate per workflow type into a single agent health dashboard, with SLO-based alerting and runbook links for each metric. **Owner:** SRE / On-Call. **Done when:** The agent health dashboard is queryable within the same observability platform as service latency and error rate, with a configured alert threshold and a runbook for each metric.

Agentic Context Graph Data Sovereignty Architecture

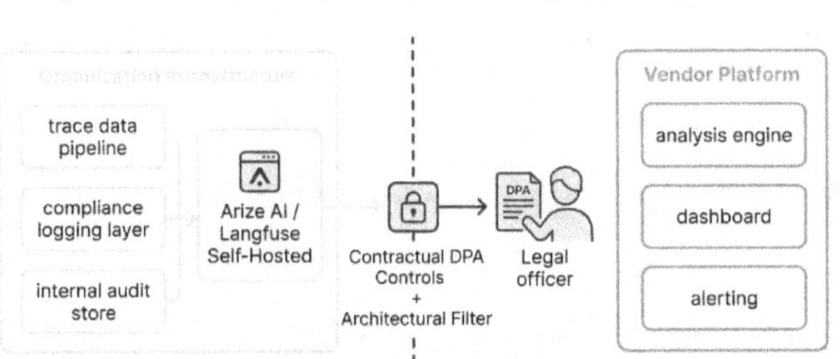

Trace data must pass through organizational compliance controls before vendor processing — implicit data sovereignty transfers are governance failures.

Agentic observability stack: trace instrumentation layer feeding the trace store, behavioral monitors and cost sentinels reading from the live span stream, agent health dashboard, and regression eval suite reading from aggregated telemetry

8.7 Team Conversation Starters

- If a task produces a wrong output and we pull the trace, can we identify within ten minutes the exact tool call and exact response content that caused the failure? If not, what span attributes are we missing?
- What is our current cost-per-task p99 for each production workflow type, and at what multiple of p99 does an alert fire? Has that threshold ever been tested with a synthetic loop scenario in staging?
- When our orchestrator delegates to a worker agent, does the worker's trace appear as a child of the orchestrator's delegation span in the unified trace view, or are they disconnected records that require manual correlation?

- For each tool in our agent's authorized toolset, do we have a registered output schema, an automated validation step in the instrumentation layer, and a documented circuit-breaker behavior for consecutive validation failures?
- Have we defined golden trajectories for at least the three highest-volume workflow types, and are those trajectories running as regression evals in CI? When did we last run a regression eval against production traces?
- What does our kill-switch test look like in staging? Have we verified that the graceful shutdown sequence completes correctly and the kill-event span appears in the trace with the correct trigger reason?

8.8 Key Takeaway

Agentic AI systems fail in ways that are behaviorally normal from the runtime's perspective but operationally catastrophic from the platform team's perspective. A loop that burns thousands of dollars generates no exceptions and no error-rate alerts; it just runs. Detecting and containing these failure modes requires instrumentation at a depth that standard observability stacks do not provide by default: typed span hierarchies with tool-call input and output capture, schema validation attributes, span linking across agent boundaries, and behavioral monitors that operate on the pattern of tool calls over time rather than the result of any individual call. The platform team that builds this instrumentation layer before going to production owns a system it can debug, govern, and improve at the speed of incidents. The team that skips it is operating a black box that charges by the minute.

9 AI Observability Platform Vendor Comparison and Selection Guide

9.1 Opening Scenario

Maya Okonkwo had run platform PoCs before and was familiar with the failure modes. When her ML platform team kicked off a six-week vendor PoC for AI observability, she wrote the acceptance criteria document the night before the kickoff call and shared it with all three vendors upfront.

The criteria were specific. OTel-native ingestion: the platform must accept OpenTelemetry spans without any proprietary SDK at the call site. Agent-trace support: multi-hop agent spans must arrive intact and render as a linked trace timeline, not a flat event list. Eval harness API: a REST or gRPC endpoint must accept (input, output, ground-truth) triples and return structured scores with run metadata. Custom-metric SDK: the team must define and push arbitrary-named metrics entirely via the API. Audit-log immutability: evaluation results, alert fires, and config changes must be written to an append-only log with no delete capability, accessible only via a signed break-glass procedure.

Three vendors made the shortlist. All three passed the demo round. Vendor A had the most polished dashboard. Vendor B had the deepest LLM evaluation story. Vendor C had the strongest enterprise sales motion and the most familiar brand.

Week three, Maya's team ran the OTel ingestion test. They instrumented a minimal inference service with the OTel Python SDK and GenAI semantic conventions, configured the OTLP exporter to point to the vendor collector endpoint, and monitored the trace stream. Vendor A's collector silently dropped every attribute whose key began with gen_ai. The spans arrived, but the AI-specific payload was gone. The vendor's collector had no support for the GenAI semantic convention namespace; the OTel path was a roadmap item.

Vendor A lost the PoC that afternoon. The acceptance criteria document, written the night before, saved the team from a contract that would have required re-instrumenting the entire inference fleet.

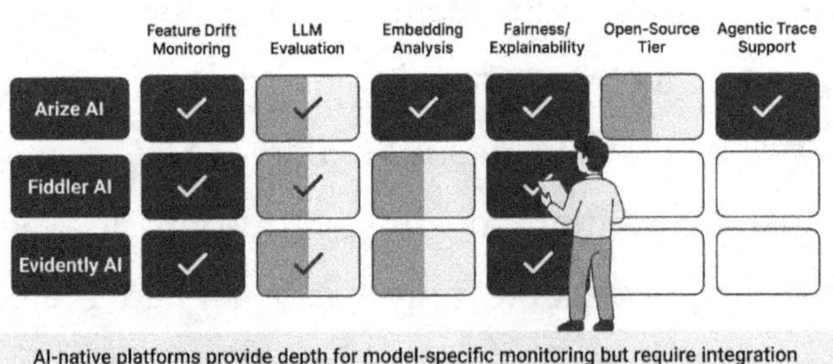

Six-week vendor PoC timeline with acceptance criteria gates and elimination decision points

9.2 Why It Matters for Production Systems

Vendor selection failures in AI observability follow a recognizable pattern: the evaluation team runs a demo, is impressed by the dashboard, and signs a contract. Six to eighteen months later, the team discovers that the platform cannot handle the actual telemetry semantics their workload produces, that instrumentation requires a proprietary agent, creating lock-in, or that governance features sit behind a higher pricing tier. Re-platforming costs engineering time the team cannot spare and sometimes requires re-instrumenting every model service in the portfolio. The root cause is evaluating the wrong things: teams over-index on dashboard aesthetics and pre-configured alert templates, which are cosmetic, and under-index on telemetry semantics, instrumentation architecture, and lock-in risk, which are structural.

Every observability platform has an internal data model that determines what telemetry it can ingest and what queries become possible downstream. AI workloads add concepts that general-purpose APM platforms have not modeled well. An LLM inference call includes a prompt template, rendered prompt text, a model identifier, token counts, a finish reason, and, optionally, a quality score from an evaluator-as-judge pipeline. An agent execution carries a root span for the task, child spans for each tool call, and span links connecting retrieved documents to the retrieval span. If the platform cannot represent these concepts natively, it drops the attributes, stores them as opaque blobs, or requires a transformation layer that recreates lock-in. The correct evaluation question is not "can this platform ingest traces?" but "can it ingest GenAI semantic convention spans without loss and execute a span-level query filtering on gen_ai.system, gen_ai.request.model, and gen_ai.usage.input_tokens simultaneously?"

Instrumentation cost is the engineering labor required to connect a model service to the observability platform and keep that connection working across framework upgrades. Low cost: install the OTel SDK, configure the OTLP exporter, and set the endpoint. High cost: install a proprietary agent, annotate inference functions with vendor-specific decorators, and handle breaking changes on the vendor's release schedule for every language runtime in the fleet. A team with fifty model services deploying new versions weekly cannot absorb the instrumentation overhead that is acceptable for ten services. OTel-native instrumentation inverts the dynamic: switching vendors becomes a change to the collector configuration, not a code change in the inference service. Lock-in operates at two layers: instrumentation lock-in is the proprietary SDK in the inference code; schema lock-in is proprietary telemetry storage with no machine-readable export. A defensible architecture minimizes both through OTel-native collection and a required schema export test during the PoC.

APM Platforms With AI Modules: Datadog, New Relic, Dynatrace Coverage Map

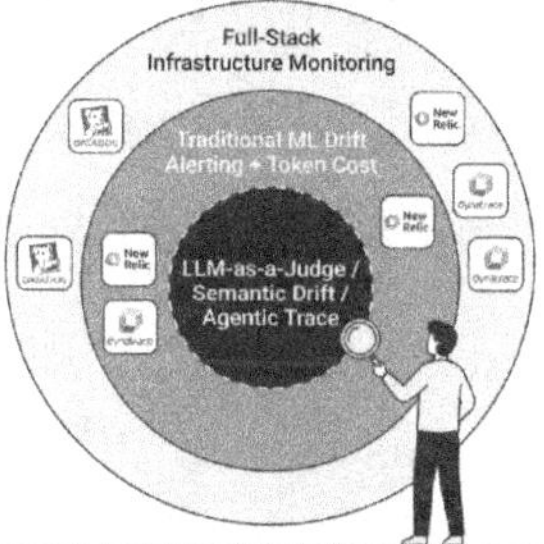

APM AI modules cover infrastructure and basic drift monitoring well; verify explicitly whether session-level agentic evaluation meets your requirements.

Lock-in topology: instrumentation layer, schema layer, and workflow layer with escape hatches at each level

9.3 Capability Axes That Actually Matter for Engineering Teams

Feature lists are marketing documents. The engineering team needs to test specific, measurable capabilities against actual production telemetry. Each axis below has a concrete test to run during the PoC.

OTel-Native Ingestion and Span Model Fidelity

OTel-native ingestion means the platform receives spans over OTLP, deserializes them without loss, and makes all span attributes queryable without any proprietary agent on the emit side. The test: instrument a minimal inference service with only the OTel SDK and GenAI semantic convention attributes, point the OTLP exporter at the platform's collector, and verify every gen_ai.* attribute arrives intact and is filterable. Platforms that strip unknown attribute namespaces fail this test regardless of marketing copy. Agent workloads require the platform to preserve parent-child relationships across span trees, handle out-of-order span arrival, and support SpanLink semantics. Platforms that flatten agent traces into event lists cannot support root-cause analysis for multi-hop agent failures. General APM platforms have mature OTLP ingestion but variable GenAI attribute coverage. Arize Phoenix is the most OTel-aligned option in the ML-native tier; in the LLM-native tier, Langfuse accepts OTLP directly.

Custom Metric SDKs and Eval Harness Primitives

Every production LLM application needs metrics that no vendor anticipates: policy-violation rate by content category, retrieval precision at K, and goal-completion rate, scored on weekly samples. The custom metric SDK test: define a named metric with arbitrary dimensions, push values from the eval pipeline via API, and query that metric alongside built-in metrics without admin UI registration. The eval harness API test: submit an (input, output, expected) triple to the platform's evaluation endpoint, receive a structured score object with evaluator name, score value, reasoning text, and run ID, and retrieve that object via API without the UI. Arize and Arize Phoenix have strong eval primitives, including LLM-as-a-judge evaluators configured via template and Python scorer functions. Langfuse accepts arbitrary numeric and categorical scores via a scoring API. Braintrust leads the LLM-native tier in eval depth, with experiment tracking and regression detection across prompt and model versions.

Tiers 3 and 4: LLMOps Tools and Data Observability Platform Positioning

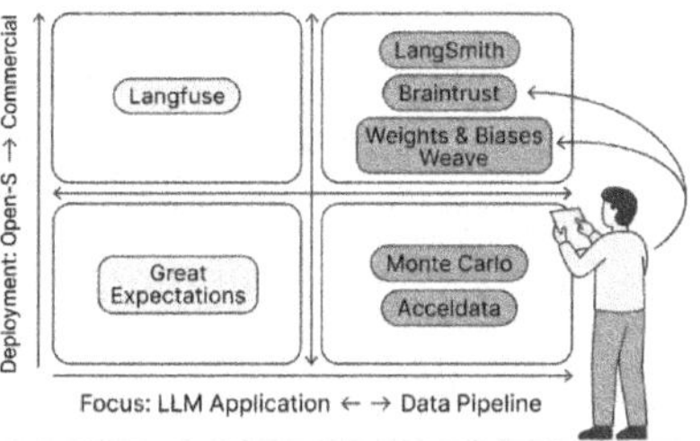

Tier 3 and Tier 4 tools complement rather than compete: LLM-specific and data observability platforms each monitor different layers of the AI system stack.

Eval harness API flow: inference spans arrive, eval pipeline scores outputs, scores pushed to platform, unified query across spans and scores

Audit and Attestation APIs, Governance, RBAC, SSO, and Data Residency

For models in regulated contexts, audit-log immutability is a hard requirement: the platform must maintain an append-only record of evaluation results, alert fires, configuration changes, and human-review events via the API, with no delete capability at the operator role level. The audit API test: retrieve a page of events, verify each carries a cryptographic event ID, millisecond-precision timestamp, actor identity, event type, and before/after payload, then attempt a delete via any available API call. Any delete path accessible without a signed break-glass procedure fails. Governance integration means the platform emits evaluation summaries and audit events to external systems via a webhook or event stream in structured JSON format. RBAC must cover read-only analyst roles, write roles scoped to a project, admin roles, and service account roles for pipelines. SSO via SAML 2.0 or OIDC is a baseline enterprise requirement. Arize Phoenix and Langfuse are both fully self-hostable. BYOK and customer-managed keys address the case where the vendor hosts data, but the organization controls encryption. FIPS 140-2 compliance is required for US federal contractors and many financial institutions; very few AI observability platforms have FIPS-validated modules.

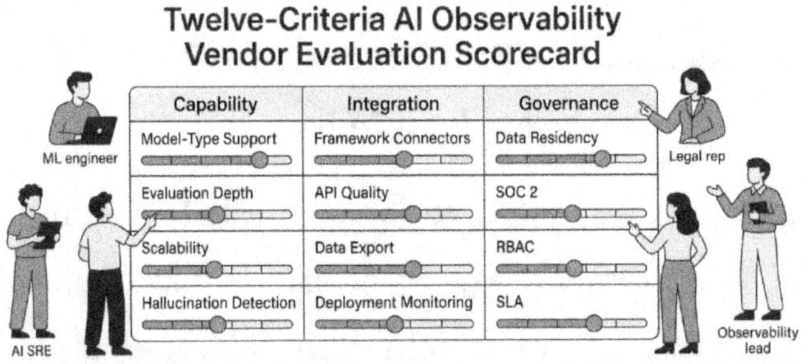

Apply a weighted twelve-criteria scorecard: regulated deployments must place the highest weight on **governance criteria.**

Data residency options: SaaS with regional selection, BYOK, self-hosted, and airgap deployment mapped to compliance profiles

9.4 Platform Categories as a Taxonomy, Not a Leaderboard

The AI observability market is not a ranked list. It is a set of distinct categories, each optimized for a different primary use case, and selecting a platform requires identifying which category fits the workload.

General APM Extensions: Datadog, Grafana, New Relic

General APM platforms entered AI observability by adding AI-specific modules to existing full-stack monitoring products. The primary strength is integration breadth: a team using Datadog for API service SLOs and on-call alerting can route LLM traces into the same platform without new infrastructure, and correlation between model latency and downstream service impact is available because everything shares the same trace store. Datadog's LLM Observability product preserves most GenAI convention attributes and supports prompt-level trace visualization. Grafana with Tempo accepts arbitrary OTLP spans without attribute loss and is the most OTel-native backend in this tier. New Relic's AI Monitoring is strong on infrastructure correlation but limited on structured evaluation pipelines. The primary gap is in evaluation depth: APM platforms assume correct behavior is deterministic, whereas LLM quality assessment requires semantic judgment about response accuracy and adherence to content policies. The

OTel instrumentation posture is the strongest among all tiers.

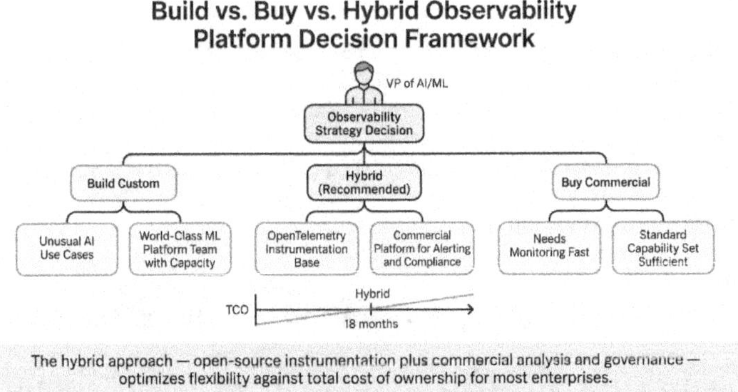

The hybrid approach — open-source instrumentation plus commercial analysis and governance — optimizes flexibility against total cost of ownership for most enterprises.

APM extension capability coverage: infrastructure correlation strength, evaluation depth gap, and OTel posture across Datadog, Grafana, and New Relic

ML-Native Platforms: Arize, Fiddler, WhyLabs, Evidently

ML-native platforms were built from the ground up to monitor AI systems, with data models centered on model inputs, outputs, and ground-truth labels, and evaluation primitives covering drift, feature importance, calibration, and fairness. Arize AI is the most comprehensive for teams running a mix of traditional ML and LLM workloads; Arize Phoenix is the reference self-hosted option, accepting OTLP spans with the most complete GenAI semantic convention support of any platform in this tier. Fiddler AI is strongest in regulated industries, with SHAP-based feature attribution and fairness monitoring for credit scoring and healthcare risk models. WhyLabs generates statistical sketches of inputs and outputs at low overhead via the WhyLogs library. Evidently, AI has the strongest open-source footprint, producing drift and LLM evaluation reports as structured JSON; teams can graduate to Evidently Cloud for governance features without changing evaluation logic.

LLM-Native Platforms: Langfuse, Arize Phoenix, LangSmith, Braintrust, Helicone

LLM-native platforms are purpose-built for LLM applications, RAG pipelines, and multi-agent systems. Langfuse is the open-source leader: trace ingestion via SDK or OTLP, a scoring API for attaching evaluation results to traces, prompt management for versioned templates, and dataset management for golden sets, all self-hosted with no runtime dependency on Langfuse cloud infrastructure. LangSmith provides seamless LangChain ecosystem integration but assumes the LangChain tracing interface rather than OTLP, recreating instrumentation lock-in for teams outside that ecosystem. Braintrust focuses on evaluation-first workflows with custom Python scorers, configurable LLM-as-a-judge rubrics, and pairwise comparison experiments. Helicone operates as a transparent AI gateway proxy, delivering low instrumentation cost at the cost of becoming a critical path dependency for inference traffic. Arize Phoenix's dual role as a standalone open-source platform and the foundation for Arize's hosted product makes it the reference implementation every PoC should include.

Open-Source and Self-Hosted Stacks

The open-source AI observability stack is composable: OTel Collector as telemetry routing backbone, Grafana Tempo as trace backend, Prometheus or Mimir for metrics, Loki for logs, Grafana for visualization, and Arize Phoenix or Evidently for AI-specific evaluation. Self-hosting is the right architecture when data residency requirements eliminate commercial options or when the ML platform team has dedicated headcount for infrastructure. Outside these conditions, maintenance overhead typically exceeds the cost of commercial licensing. The sidecar collector pattern is the production-grade deployment model. Each inference pod runs an OTel Collector sidecar that receives OTLP over localhost, applies sampling and attribute filtering, and forwards to the backend. eBPF auto-instrumentation supplements SDK-based instrumentation for system-level telemetry, but does not reach Python-level LLM SDK calls.

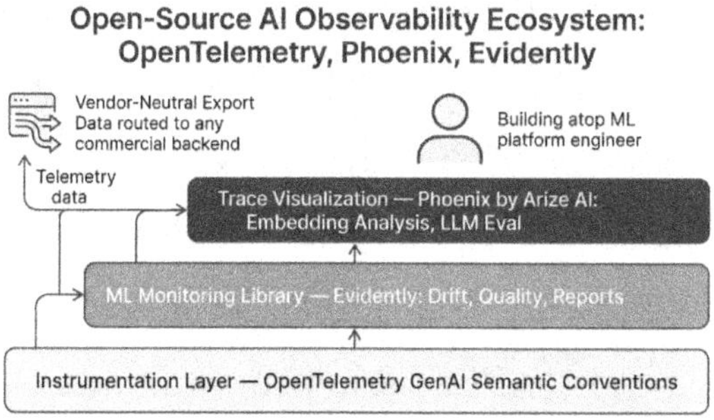

OpenTelemetry instrumentation is the architectural keystone: it preserves vendor optionality and prevents lock-in at the data collection layer.

Self-hosted OTel collector pipeline: sidecar per inference pod, fan-out to trace and metrics backends, eval pipeline as sidecar process.

9.5 Running a Rigorous PoC: Acceptance Criteria, Scoring, and the Build-vs-Buy Decision

A vendor PoC that does not test actual production telemetry shape is not a PoC. The evaluation team must instrument the real inference workload and run it against real telemetry.

Defining Acceptance Criteria and the Scoring Rubric

Acceptance criteria should be binary tests with clear pass/fail outcomes that any team member can replicate independently. "OTel ingestion works" is not a binary test; "OTLP spans with gen_ai.* attributes arrive intact and are filterable within 30 seconds" is. Share the criteria document with all vendors at kickoff: vendors who know them upfront will either pass or proactively disclose gaps, and vendors who claim capabilities they cannot demonstrate have failed a vendor-integrity test. A minimal set of acceptance criteria covers seven binary tests: OTel-native ingestion without gen_ai.* attribute loss; agent trace fidelity across at least three span levels with span links rendered as a linked timeline; eval harness API usable programmatically without the UI; custom metric push and query entirely via API; audit log immutability with no delete API below a break-glass procedure; schema export in a documented machine-readable format; and replay export of prompt-response pairs for offline evaluation. Secondary criteria, such as dashboard usability and ecosystem health, should be scored on a weighted scale, but none should block a candidate that passes all seven binary tests.

Instrumentation Cost, Switching Cost, and Multi-Vendor Architectures

Instrumentation cost should be measured during the PoC rather than estimated: how many lines of inference-service code change, and how long until first-trace visibility? A platform requiring only OTel collector configuration has near-zero instrumentation cost. A platform requiring vendor SDK installation and function annotation incurs high costs that compound with every new service or framework upgrade. Switching cost has two components: instrumentation switching cost, which is zero with OTel-native collection and no proprietary SDK, and historical data switching cost, which is bounded with a documented schema export and unbounded without one. Multi-vendor architectures are a valid production posture: a general APM platform handles infrastructure SLO alerting. In contrast, a dedicated LLM observability platform handles trace visualization and evaluation, with the OTel Collector fanning telemetry to both backends simultaneously. Explicit alert ownership boundaries ensure that both teams know which runbook to follow when the same incident triggers alerts for both.

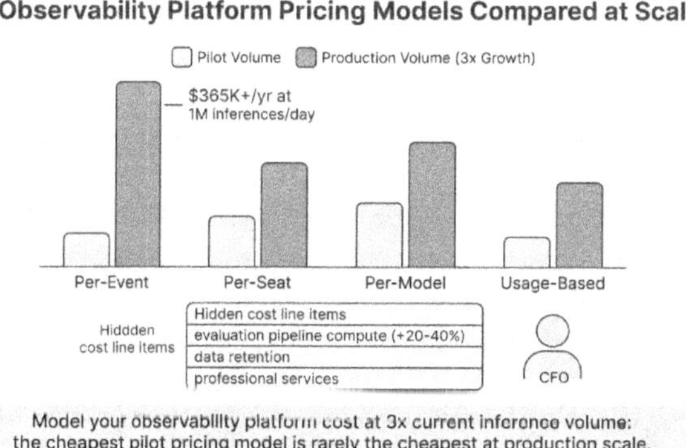

Model your observability platform cost at 3x current inference volume: the cheapest pilot pricing model is rarely the cheapest at production scale.

Multi-vendor observability architecture: OTel Collector fan-out to APM backend, LLM trace backend, and data quality backend with alert ownership boundaries

Build vs. Buy: A Decision Frame for Architecture Teams

A team that chooses to build can assemble a defensible production stack from Arize Phoenix, Evidently, the OTel Collector, and Grafana at zero licensing cost. The costs that complicate this are operational: self-hosted infrastructure requires a team that can configure it, upgrade it, debug it at 2 AM, and evolve it as requirements change. For a team also responsible for inference infrastructure and the model deployment pipeline, adding an observability stack is a real cost. The build decision is strongest when data residency requirements eliminate all commercial options, or when dedicated infrastructure headcount exceeds the inference platform's needs. Per-event pricing, common in the LLM-native tier, can produce unexpected costs at scale. At $0.002 per event and 2 million LLM calls per day, the annual observability cost reaches seven figures before evaluation pipeline compute. The hybrid architecture that works for most teams applies OTel-native instrumentation everywhere and buys commercial backends for specific capability layers: the instrumentation layer carries zero vendor dependency, the trace and metrics storage layer uses a commercial platform, and the eval harness may use a separate tool. Each layer is independently replaceable without touching the inference code.

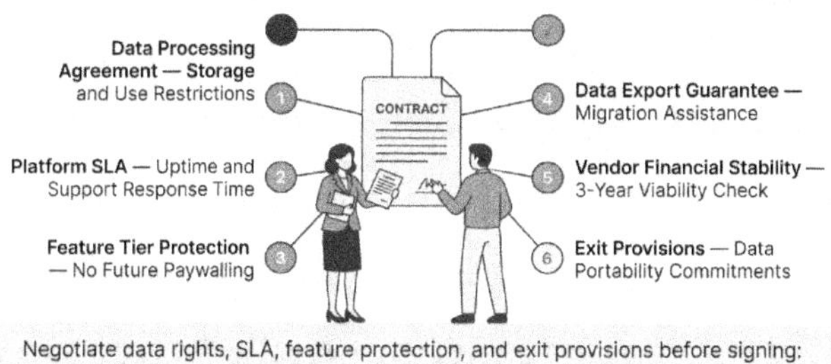

Negotiate data rights, SLA, feature protection, and exit provisions before signing: legal review of vendor terms is not optional for production AI systems.

Build-vs-buy cost comparison: instrumentation cost, operational overhead, licensing, and switching cost across inference volume tiers

9.6 Technical Team Checklist: Vendor Evaluation

- **Action:** Author a binary acceptance criteria document covering OTel-native ingestion, agent-trace span fidelity, eval harness API, custom metric push, audit log immutability, schema export, and replay export. Share it with all PoC vendors at kickoff. **Owner:** Architect. **Done when:** Document is version-controlled and confirmed received by each vendor before PoC start.
- **Action:** Build a minimal inference fixture using only the OTel SDK with GenAI semantic convention attributes and no vendor SDK. Point the OTLP exporter at each finalist's collector endpoint and record attribute fidelity using a structured diff of emitted versus received span attributes. **Owner:** ML Platform Eng. **Done when:** Attribute fidelity test results are documented for all

finalists, and attribute loss is reported to the vendor with a request for a remediation timeline.

- **Action:** Submit a multi-hop agent trace with at least three span levels and at least one span link to each finalist and document whether the platform renders a linked span tree or a flat event list. **Owner:** ML Platform Eng. **Done when:** Agent trace rendering result is captured as a screenshot and a written comparison in the evaluation document.
- **Action:** Call the eval harness API of each finalist programmatically (no UI) to submit a scored evaluation triple and retrieve the resulting score object. Verify the schema includes evaluator name, score value, reasoning text, and evaluation run ID. **Owner:** ML Engineer. **Done when:** the API call is documented with request and response payloads that confirm the schema satisfies the requirements.
- **Action:** Export a one-week window of traces and metrics from each finalist in a machine-readable format without vendor tooling assistance. Parse the export and compare the column schema against the documented export schema. **Owner:** Data Eng. **Done when:** Export is successfully parsed, and schema comparison is documented.
- **Action:** Test audit log API for each finalist: retrieve events, verify required fields, and attempt a delete or

modify API call at the highest non-break-glass permission level. **Owner:** Security Eng. **Done when:** Audit log fields are verified, and any accessible delete capability is flagged as an acceptance criteria failure.

- **Action:** Model the observability platform cost at three times the current inference volume, including base licensing, evaluation pipeline compute, data retention, and integration engineering. Compare against the estimated build cost for an equivalent self-hosted stack. **Owner:** ML Platform Eng. **Done when:** The cost model is reviewed by the architecture team, and the build-vs-buy decision is documented with assumptions visible.
- **Action:** Verify SSO integration (SAML 2.0 or OIDC) and RBAC role model against the team's identity provider. For regulated deployments, verify the status of the FIPS cryptographic module and the data-residency options before the PoC begins. **Owner:** Security Eng. **Done when:** SSO login is demonstrated against the production identity provider, and RBAC role assignments are tested for each required role type.
- **Action:** Configure a sidecar OTel Collector deployment for the PoC inference service and validate that collector-side sampling and attribute filtering operate correctly before telemetry reaches the vendor backend. **Owner:** ML Platform Eng. **Done when:** Sampling configuration is

validated at the collector level and confirmed not to affect inference service behavior.

- **Action:** Review the vendor contract for data processing agreement terms, SLA commitments, feature tier protections against pay-walling, and data export guarantees. Flag any clause requiring proprietary SDK use as a contract condition. **Owner:** Architect. **Done when:** Legal review is complete, and any flagged terms are resolved or accepted with documented risk acceptance.

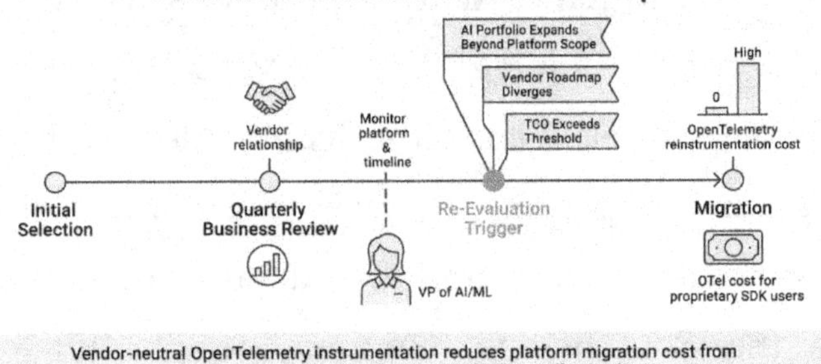

PoC acceptance criteria evaluation matrix: capability axes across rows, finalist vendors across columns, pass/fail/partial results

9.7 Team Conversation Starters

- If the vendor's OTel ingestion path loses any gen_ai.* attributes at the collector boundary, how many inference services need re-instrumentation with the vendor's proprietary SDK as a workaround, and what does that migration timeline look like against the current release schedule?

- When the eval harness cannot be called programmatically without going through the vendor's UI, what does that mean for automating evaluation gating in the CI/CD pipeline and for reproducing evaluation results during an incident review?
- If the platform does not support schema export in a documented machine-readable format, how would the team reconstruct six months of evaluation history after a platform migration, and who owns that reconstruction cost?
- For the multi-vendor architecture under consideration, who owns the alert routing configuration enforcing the boundary between infrastructure SLO alerts and model quality alerts, and what is the runbook when those alerts conflict on the same incident?
- At three times the current inference volume, does the selected platform's per-event pricing remain within budget, and is there a contract mechanism to convert to a flat-rate tier if volume crosses a defined threshold?
- When the vendor changes pricing tiers or is acquired in year two of the contract, is the team's OTel-native instrumentation posture sufficient to make switching cost bounded, and has the switching cost been measured during the PoC rather than estimated?

9.8 Key Takeaway

The difference between a vendor selection that holds for three years and one that ends in re-platforming is the quality of the acceptance criteria written before the PoC starts. Platforms that look identical in demos diverge sharply on OTel ingestion fidelity, agent-trace span model, eval harness API quality, and audit log immutability when tested against real production telemetry. No category is universally correct: general APM extensions win on infrastructure integration and operational familiarity; ML-native platforms win on evaluation depth for traditional model workloads; LLM-native platforms win on trace fidelity and prompt management for LLM applications; and self-hosted stacks win on data residency and customization at the cost of operational overhead. The architectural decision that protects every choice is OTel-native instrumentation at the call site, making the observability backend a pluggable configuration concern rather than a permanent commitment.

10 Building Your Enterprise AI Observability Architecture

10.1 Opening Scenario

Maya Okonkwo inherited the problem on her first day as principal architect on the ML platform team. The engineering manager handed her a spreadsheet with fourteen rows, one per production model, and a column labeled "monitoring tool" listing seven different answers: Prometheus with custom exporters, a vendor-managed SaaS dashboard, raw logs written to S3, a home-built Flask app that polled a model endpoint every five minutes, Datadog APM with zero ML-specific instrumentation, a Grafana instance last updated eight months ago, and for two models, nothing at all.

She pulled the platform team into a conference room for three hours at a whiteboard. She drew four horizontal swim lanes: telemetry emission at the model server, collection and schema validation, tiered storage, and visualization plus alerting. Three teams propagated no trace context across inference calls. Two teams wrote prediction outputs to logs with no schema contract. One team had configured Prometheus to scrape an endpoint that returned stale cache data, resulting in readings that lagged reality by up to 4 minutes under load.

The session surfaced a shared problem: all 14 teams were independently reinventing the same OTel collector configuration. The collector layer was the logical starting point because it could sit in front of every existing setup as a thin forwarding layer, without requiring teams to rewrite their model-serving code. Maya tasked two engineers with a reference collector configuration any team could adopt in an afternoon: handle attribute validation, sampling policy, and downstream routing, letting each team keep existing exporters as secondary sinks while the collector forwarded a canonical event stream to the shared ingestion pipeline. That prototype became the foundation of the reference architecture, and the platform team shipped six weeks later.

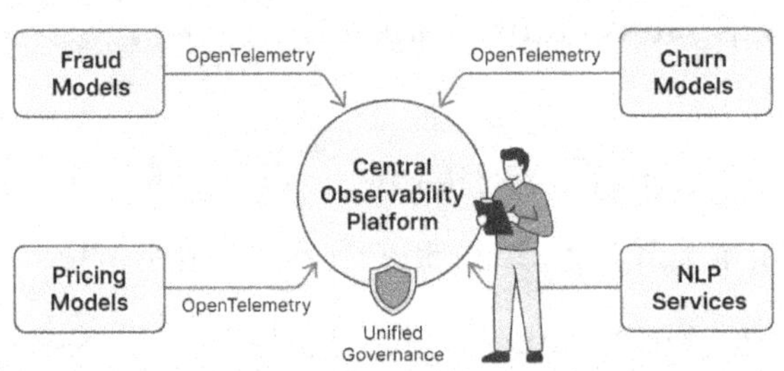

Whiteboard sketch of the four-layer reference architecture: emission, collection, tiered storage, and visualization

10.2 Why It Matters for Production Systems

When every model team builds its own observability stack, the organization ends up with incompatible telemetry schemas, divergent retention policies, and alerting rules that only the originating team can interpret. That fragmentation compounds as the portfolio grows, and the cost surfaces in three distinct places.

The first is incident response. When a production model starts returning anomalous outputs, the on-call engineer needs to correlate inference latency, input feature distributions, model output distributions, and upstream data pipeline health within a single investigation window. If those signals live in four systems with different query interfaces, MTTD stretches from minutes into hours, and the fragmented setup makes incidents harder to detect in the first place because no single alert configuration has visibility across the full signal space. The second is model promotion: a team gating a new model version on production-comparable evaluation metrics needs the production observability store as a reference distribution; without a standard query interface, that gate becomes a manual process of downloading CSVs and running local scripts. The third is governance: assembling evidence of continuous model monitoring from fourteen fragmented stores is a multi-week manual effort for organizations subject to SR 11-

7, the EU AI Act model risk provisions, or internal model risk management frameworks.

A reference architecture addresses these problems by enforcing a small set of constraints early: a shared semantic convention for telemetry attributes, a standard OTel collector configuration, a tiered storage design separating hot operational data from warm analytical data from cold audit archives, and integration points where the CI/CD pipeline, model registry, and feature store emit lifecycle events into the shared telemetry stream. With this foundation, a new team can instrument their model server in an afternoon, tooling investment is amortized across every model in the portfolio, and drift detectors can share a reference window rather than each team building separate detection logic against its own schema.

10.3 Reference Architecture Components: Telemetry Collectors, Storage Layers, and the Control Plane

A production AI observability reference architecture is a composition of purpose-built components: the OTel collector tier, the ingestion pipeline, the hot-tier time-series and trace stores, the warm-tier analytical layer, the cold-tier audit archive, the eval store, the model registry integration, the feature store integration, the dashboard layer, the alerting layer, and the governance attestation pipeline. Teams that jump to vendor selection without this component decomposition end up with overlapping tools and critical gaps.

The Telemetry Emission and Collection Tier

The ML platform engineer deploys an OpenTelemetry SDK in each model-serving container to wrap the inference handler. The SDK emits a span for each inference request, attaching the model version, the input feature vector hash, the output value, the serving latency, and the post-processing steps. For generative AI endpoints, the span also carries the prompt template identifier, token counts, temperature, and the full response, where privacy permits. The OTel collector sits one layer above the SDK, deploying as a DaemonSet in Kubernetes or as a per-pod sidecar where blast-radius isolation is required. The collector validates incoming spans against the schema registry, applies the sampling policy, and routes events to downstream sinks.

Sampling strategy is the configuration decision with the largest impact on cost and observability fidelity. Head-based sampling decides at the start of a request before any downstream spans are emitted; it is cheap but statistically blind, discarding the rare high-latency outliers that matter most for diagnosis. Tail-based sampling buffers all spans until the root span completes, then retains all traces that contain errors or latency anomalies. Tail-based sampling is the correct default for AI inference workloads because the failure modes (hallucinated outputs, distribution shifts, retrieval failures) are precisely the rare events that head-based sampling discards.

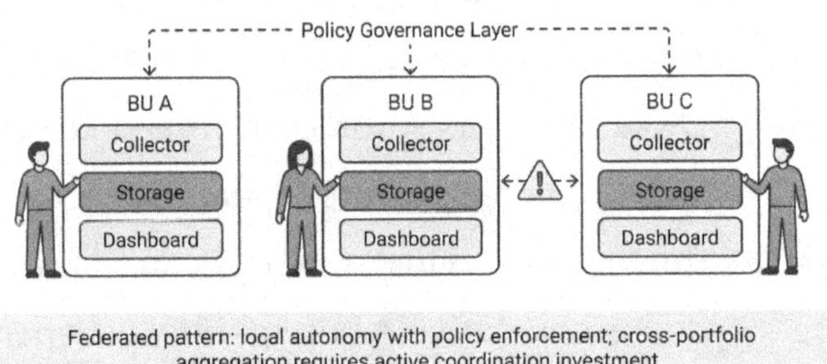

Federated pattern: local autonomy with policy enforcement; cross-portfolio aggregation requires active coordination investment.

OTel collector deployment: DaemonSet node agent with tail-based sampling decision point and schema validation at the collector boundary

Storage Tiers: Hot, Warm, Cold, and the Eval Store

Enterprise AI observability generates data types with fundamentally different velocity and retention requirements. The hot tier handles real-time operational data: a time-series database for metrics, a distributed tracing backend for spans (Tempo or Jaeger), and an indexed log store, with seven to thirty-day retention before downsampling and forwarding to the warm tier. The warm tier holds aggregated metrics and analytical data: PSI scores, KS-statistic results, feature distribution histograms, and model performance metric time series. Apache Iceberg on object storage with Databricks or

Snowflake as the query engine is the standard pattern here because Iceberg's schema evolution support allows the platform team to add new metric columns without rewriting existing data or breaking existing dashboard queries. The cold tier is the immutable audit archive: every event must be signed and locked, and the cold-tier write path must be separate from the operational write path to prevent an ingestion pipeline outage from creating audit gaps. The eval store sits adjacent to this three-tier model, holding golden dataset results, evaluator-as-judge scores, human review labels, and ground truth labels, with its schema versioned to record the evaluator version alongside each score.

Hub-and-Spoke Telemetry Architecture

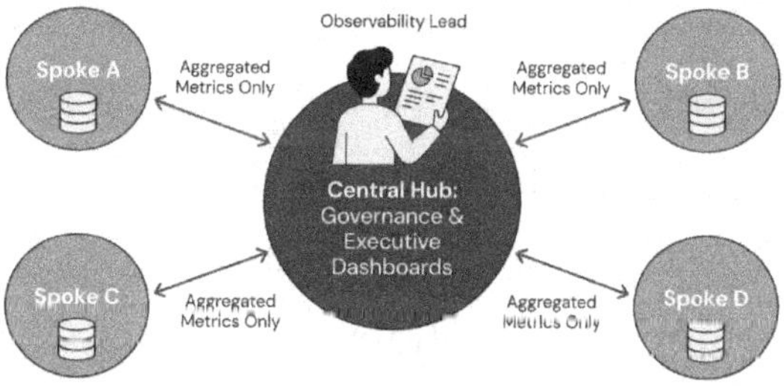

Hub-and-spoke: local operations at spokes, portfolio governance and audit reporting centralized at hub.

Three-tier storage layout: hot operational stores, warm Iceberg analytical layer, cold immutable archive, with eval store feeding the promotion gate

Model Registry, Feature Store, and Dashboard Layer

Every model in production should have a corresponding model registry entry recording the model version, training dataset version, staging evaluation metrics, promotion timestamp, and the approving engineer. The observability platform consumes model registry events via webhook to keep its inventory synchronized. Feature store integration is the highest-leverage upstream hook: when an upstream data pipeline changes and the feature store begins serving a different statistical distribution, the observability platform detects the shift within minutes, before model output metrics have moved enough to trigger a model-level alert.

The dashboard layer maintains two distinct interfaces on the same warm-tier storage. The operational dashboard is a real-time instrument for ML engineers and SREs: current inference latency percentiles, recent PSI and KS-statistic values, data quality check pass rates, and active alert counts. The portfolio dashboard is a risk-tracking instrument for the platform team: how many models are within their performance SLO, drift severity trends over the past ninety days, models approaching retraining triggers, and whether any compliance-relevant alert was escalated to human review in the past week.

10.4 Build Patterns: Tenancy Models, Instrumentation, and Retention Policy

Knowing what components belong in the reference architecture answers the structural question. Knowing how to configure and deploy them answers the operational question. The build patterns here cover the decisions that most directly affect long-term operability: tenancy model, instrumentation deployment strategy, push versus pull metrics, sampling configuration, and retention tier policy.

Single-Tenant vs Multi-Tenant Deployments and Federated Tenancy

The tenancy model determines how the observability infrastructure is partitioned across model teams and business units. A single-tenant deployment runs a complete stack per team, which is appropriate when data sovereignty requirements prohibit mixing telemetry, but it leads to portfolio-level fragmentation. A multi-tenant deployment runs a single shared stack for all teams, with logical separation achieved through namespace conventions, access control, and attribute-based routing: every span carries a service. name resource attribute encoding the team and model identifier, and the collector routes events to the correct storage partition.

Federated tenancy is the hybrid pattern that most large enterprises converge on. Each team's collector acts as an independent data plane handling local ingestion, schema validation, and operational alerting. In contrast, the platform team's central hub stores forwarded events in the shared warm and cold tiers and powers portfolio-level dashboards and governance reports. The critical design decision is the forwarding contract, which signals travel to the hub, at what granularity, and with what latency guarantee. Teams that forward raw inference traces quickly overwhelm the hub; the correct forwarding contract sends aggregated statistics (drift scores, performance metric time series, alert events, governance attestations) rather than raw event streams.

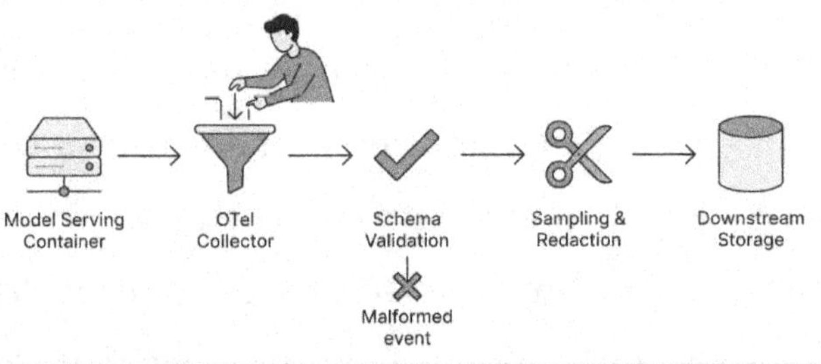

Federated tenancy topology: per-team spoke forwarding aggregated signals to the central warm-tier and cold-tier hub

Sidecar vs Gateway Instrumentation and Push vs Pull Metrics

The sidecar instrumentation pattern deploys the OTel collector as a container within the same pod as the model serving container, capturing every inference request before it leaves the pod network boundary and applying attribute enrichment (Kubernetes node name, pod IP, GPU device identifier). The gateway instrumentation pattern deploys the OTel collector as a standalone service receiving telemetry from all model serving pods over gRPC or HTTP, creating a shared resource that can bottleneck under high throughput without redundancy. The recommended production configuration combines both: a lightweight DaemonSet agent on each node for attribute enrichment and local buffering, forwarding to a gateway collector tier that handles sampling, schema validation, and downstream routing.

The push-versus-pull decision for metrics is separate from the collector deployment topology. Pull-based collection creates problems in dynamic Kubernetes environments where pods are created and destroyed faster than the scrape interval. Push-based collection, where model servers emit metric events to a collector endpoint on each update, scales better with dynamic environments and produces lower-latency data. OpenTelemetry's metrics SDK uses push-based collection by default and should be the platform-wide standard.

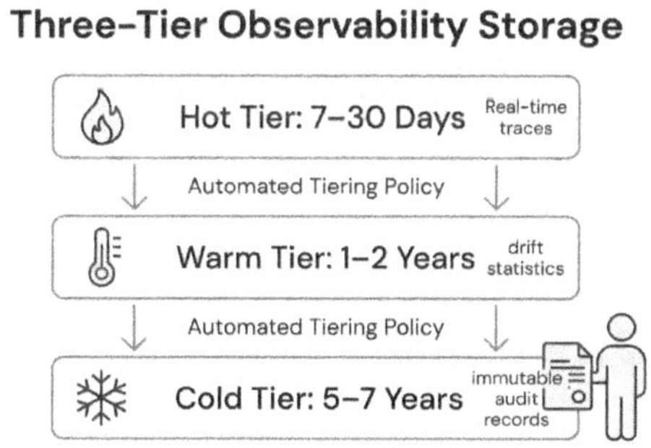

Three-tier storage maps data velocity to retention windows required by SR 11-7 and EU AI Act.

Two-tier collector pattern: DaemonSet node agent for buffering and enrichment, feeding a scalable gateway collector pool for sampling and routing

Retention Tier Design and Hot/Cold Storage Policy

Retention policy design determines both the cost of the observability infrastructure and its ability to support future investigation and governance needs. For real-time inference traces in the hot tier, the standard window is seven to thirty days, retaining only error traces and high-latency outliers beyond that. For aggregated drift statistics and performance metrics in the warm tier, a standard window of 12 to 24 months applies; this window must be reviewed whenever the organization adopts a new regulatory framework, because SR 11-7 and the EU AI Act's model risk provisions specify minimum evidence retention periods. For the cold-tier audit archive, the baseline is five to seven years, with events written with object-lock and WORM semantics and modifiable only by a privileged data lifecycle process. All tier transitions should be automated through a version-controlled lifecycle policy, reviewed annually, and referenced in the governance charter.

10.5 Integration with the Broader Platform: CI/CD Hooks, Promotion Gates, and Governance Pipelines

The reference architecture becomes qualitatively more powerful when integrated with the systems that govern the model lifecycle: the CI/CD pipeline, the data contract validation system, the governance attestation pipeline, and the on-call tooling that routes alerts to the right engineers at the right time.

CI/CD Hooks and Model Promotion Gates

The model promotion gate is the single most impactful integration point between the observability platform and the broader ML platform. The gate retrieves the production model's performance distribution from the warm-tier store, computes evaluation metrics for the candidate against the same distribution using the eval store, and checks whether the candidate's metrics fall within defined acceptance bounds. It also checks instrumentation completeness: does the model server emit the required OTel attributes, does the schema registry have an entry for the new model version, and does the collector configuration include a routing rule for the new model identifier? A promotion that passes the evaluation check but fails the instrumentation check should not go to production; a model with incomplete instrumentation is invisible from the moment it deploys. The CI/CD pipeline emits a structured promotion event carrying the model identifier, version pair, evaluation metrics, and the approving engineer; the observability platform uses this event as a baseline anchor for all subsequent drift detection and performance monitoring.

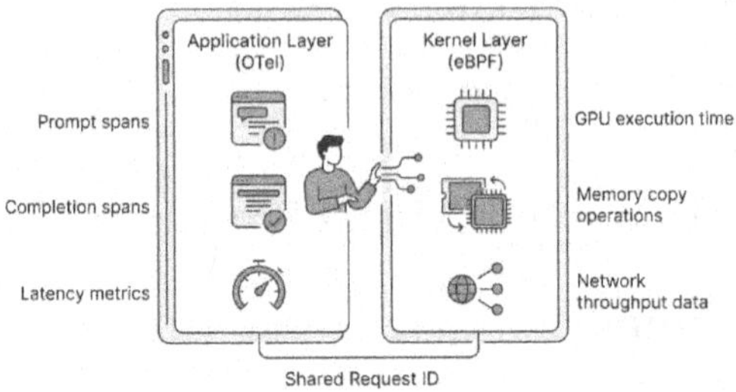

eBPF captures kernel GPU and network data without modifying model-serving application code.

Promotion gate flow: evaluation check against warm-tier reference distribution, instrumentation completeness check, and promotion event emission into the ingestion pipeline

Data Contract Validation and Feature Store Integration

A data contract is a machine-readable schema agreement between an upstream data producer and a downstream consumer. In the AI platform context, the contract between the feature store and the model server specifies feature names, data types, acceptable value ranges, and expected statistical distributions. When the upstream data pipeline produces features that violate the contract, the feature store's contract validation layer emits a violation event into the observability ingestion pipeline. A feature distribution that drifts outside the contract's bounds is a leading indicator of model output drift; instrumenting contract validation at the feature store's serving layer enables the platform team to fire alerts hours before model-level metrics indicate degradation. The data contract is a versioned artifact in the schema registry, and a schema mismatch between the candidate model's expected input and the current registry entry is a hard block on the promotion gate.

Governance Attestation Pipeline, Runbook Generation, and On-Call Integration

The governance attestation pipeline transforms observability data into compliance evidence. An attestation is a signed, timestamped record that a specific governance condition was met: the drift detector was running and healthy, model performance metrics were within SLO, feature contract validation was passing, and no unreviewed high-severity alerts were open beyond the defined SLA. The pipeline runs on a schedule, is implemented as version-controlled query templates against the warm tier, and writes each record to the cold-tier archive. When a regulator requests evidence for a specific time period, the governance engineer runs the attestation query and produces a structured evidence package from the cold-tier records.

Two-Audience Observability Dashboard Design

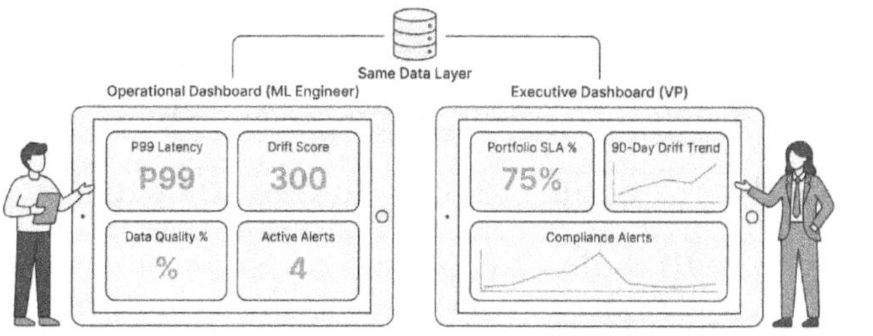

Both dashboard types draw from one data layer to prevent metric divergence within six months.

Governance attestation pipeline: warm-tier query, threshold evaluation, signed attestation record emission to cold-tier archive, and evidence package generation on demand

On-call integration routes alerts based on the model identifier, alert severity, and team ownership metadata in the model registry. High-severity alerts fire a PagerDuty or Opsgenie incident with a pre-populated runbook generated from the alert type, model identifier, and recent promotion history. Medium-severity alerts create a Slack notification and a Jira ticket. Low-severity alerts write to a review queue. The routing configuration is version-controlled and audited monthly to detect stale assignments.

Observability Integration Across the ML Lifecycle

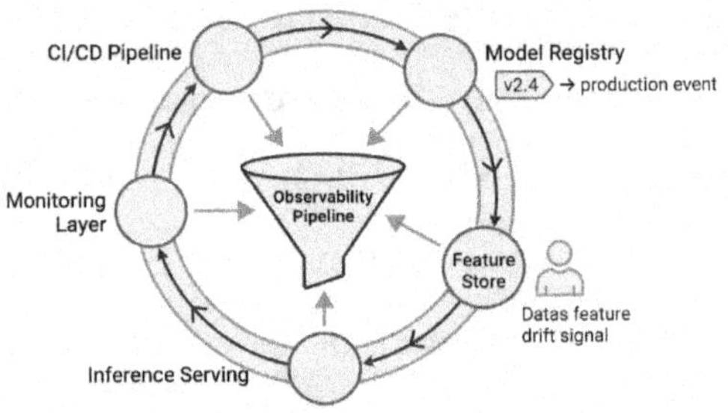

Upstream lifecycle integrations enable drift detection before model output distributions degrade.

Alert routing tiers: high-severity incident creation with runbook attachment, medium-severity Slack and Jira notification, low-severity review queue, driven by model registry ownership metadata

The complete reference architecture connects all layers through a shared ingestion pipeline: OTel spans from model servers, CI/CD promotion events, feature store contract violations, and model registry lifecycle events, all of which flow to the same ingestion endpoint. The warm tier feeds both the operational dashboards and the governance attestation pipeline. At the same time, the alerting layer reads from both tiers and routes to on-call channels based on model registry ownership metadata.

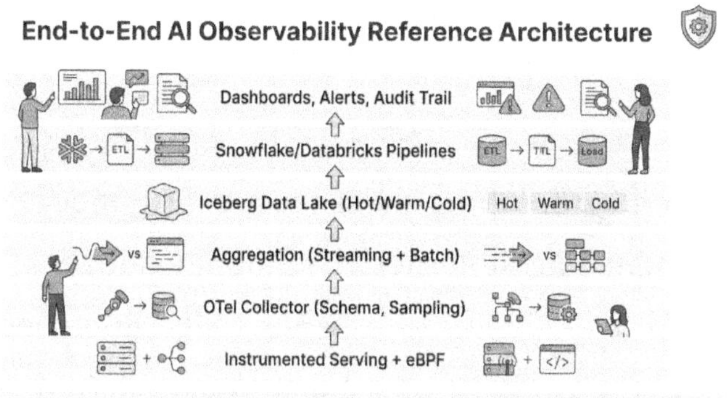

Reference architecture: instrumentation feeds governed storage that powers operational dashboards and audit evidence.

Complete reference architecture: emission tier, collector tier, hot and warm and cold storage, CI/CD gate, feature store, attestation pipeline, and on-call routing connected through the shared ingestion pipeline

10.6 Technical Team Checklist: Reference Architecture Adoption

- **Action:** Deploy a standard OTel collector as a DaemonSet on each inference cluster with tail-based sampling and schema validation at the collector boundary. **Owner:** ML Platform Eng. **Done when:** All model serving pods emit spans to the collector, schema violations are alerted, and no raw events bypass validation.

- **Action:** Establish a three-tier retention policy: hot-tier seven to thirty days, warm-tier twelve to twenty-four months, cold-tier immutable five to seven years, with automated lifecycle transitions. **Owner:** ML Platform Eng. **Done when:** Lifecycle transitions run on schedule, and the policy is version-controlled in the governance charter.
- **Action:** Instrument the CI/CD pipeline to emit a structured promotion event for every model version deployment, carrying model version, evaluation metrics, and approver identity. **Owner:** ML Engineer. **Done when:** Every promotion event is queryable in the warm-tier store within five minutes of deployment, and the drift detection baseline is automatically updated.
- **Action:** Integrate the feature store's serving layer with the OTel ingestion pipeline to emit feature distribution statistics and data contract violation events on each serving batch. **Owner:** Data Eng. **Done when:** Feature health time series are visible in the operational dashboard alongside model drift scores and contract violations fire alerts within fifteen minutes.
- **Action:** Build the governance attestation pipeline as version-controlled query templates against the warm tier, emitting signed attestation records to the cold-tier

archive daily for all tier-one production models. **Owner:** Governance Eng. **Done when:** Attestation records for all tier-one models are present in the cold-tier archive with no gaps for the past ninety days.

- **Action:** Wire the alerting layer to the on-call rotation using model-identifier-to-team-ownership mappings from the model registry, with automatic runbook attachment on high-severity incidents. **Owner:** SRE / On-Call. **Done when:** Every high-severity alert creates an incident with the correct on-call assignment and a runbook linking to relevant dashboards and the most recent promotion event.
- **Action:** Establish a federated tenancy forwarding contract specifying which aggregated signals each spoke forwards to the central hub, at what granularity, and with what latency guarantee. **Owner:** Architect. **Done when:** The forwarding contract is documented, all spokes are forwarding within agreed latency bounds, and the hub's model inventory matches the model registry.
- **Action:** Configure the promotion gate to enforce instrumentation completeness checks alongside evaluation metric checks, blocking promotions where the model's OTel configuration is incomplete or its schema registry entry is absent. **Owner:** ML Platform Eng. **Done when:** At least two promotion attempts with incomplete

instrumentation have been blocked by the gate in the past sixty days.

- **Action:** Run a monthly audit comparing the alert routing configuration against current model registry ownership records, correcting stale assignments and documenting the result. **Owner:** SRE / On-Call. **Done when:** The most recent audit shows zero stale routing assignments and is stored in the governance charter repository.

10.7 Team Conversation Starters

- When the on-call engineer gets paged for a production model degradation, which specific dashboards and query interfaces do they open, and can they reach a diagnosis without asking anyone else for access or context?
- If the feature store begins serving a feature distribution statistically different from the training distribution by a PSI of 0.25, how long does it take the observability platform to detect and alert on that shift, and what is the first thing the on-call engineer sees in the runbook?
- Which models in the production portfolio currently have instrumentation gaps, and what is the process for an engineer to discover and remediate a gap without waiting for an incident to expose it?
- When a regulator requests evidence of continuous monitoring for a specific model over the past twelve months, what is the end-to-end time from request to

evidence package delivery, and which components of the architecture are involved?

- If the shared OTel gateway collector goes down for fifteen minutes, which models lose telemetry visibility, what alerts fire, and what is the documented recovery procedure?
- How does the architecture handle a new model team that wants to emit custom metrics specific to their domain without affecting the shared schema in a way that breaks other teams' dashboards or the governance attestation queries?

10.8 Key Takeaway

A reference architecture for enterprise AI observability is the prerequisite for every other observability investment to compound rather than fragment. Its value is not in the specific tools chosen for each layer but in the enforcement of a shared telemetry contract across all model teams: a standard OTel collector configuration, a canonical attribute schema, a tiered storage design with documented retention policies, and integration points connecting the observability platform to the CI/CD pipeline, the model registry, the feature store, and the governance attestation pipeline. Teams that establish this foundation early reduce MTTD and MTTR for production incidents, enable automated promotion gates, and produce governance evidence on demand. The architecture is buildable with standard open-source and commercial components; what it requires is deliberate design decisions made early and enforced consistently as the model portfolio grows.

11 AI Alerting, Incident Response, and On-Call Escalation

11.1 Opening Scenario

The page lands at 2:47 AM on a Wednesday. Priya Subramaniam, primary on-call for the recommendations platform team, acknowledges a PagerDuty alert within ninety seconds: "rec_hitrate_p95 below SLO threshold for 18 consecutive minutes -- current 0.41, SLO floor 0.52 -- SEV-2 auto-declared." She opens her laptop and pulls the runbook.

Step one: confirm the alert is not a monitoring artifact. The raw event stream shows genuine traffic—roughly 40,000 inference calls that return lower-ranked results than the model's validation baseline. Not a fluke. Step two: inspect upstream dependencies. The feature pipeline is green. The embedding index reports a stale-refresh flag from 2:31 AM. The nightly index swap did not complete: the new candidate index loaded 60 percent of shards before a storage quota error halted the job. The serving layer mixes stale and fresh shard responses, producing an incoherent ranking.

Step three is the decision gate: "If root cause is index integrity failure, trigger kill switch KS-04 to route traffic to fallback model (deterministic popularity-weighted ranker) and page secondary on-call to repair index." She triggers KS-04 via the internal CLI -- one command, one confirmation prompt, thirty seconds. Hit rate climbs back through 0.52 within four minutes. She pages secondary on-call, documents the trigger in the incident ticket, and sets an attestation freeze on the embedding index deployment pipeline.

At 4:10 AM, she files the incident review stub, including the timeline, root cause, kill-switch action, and current state. By the time her team wakes up, the runbook worked exactly as designed, the fallback model is serving acceptable results, and there is a paper trail a regulator could read. That is the goal: a system that contains failures before they compound, not heroics in the middle of the night.

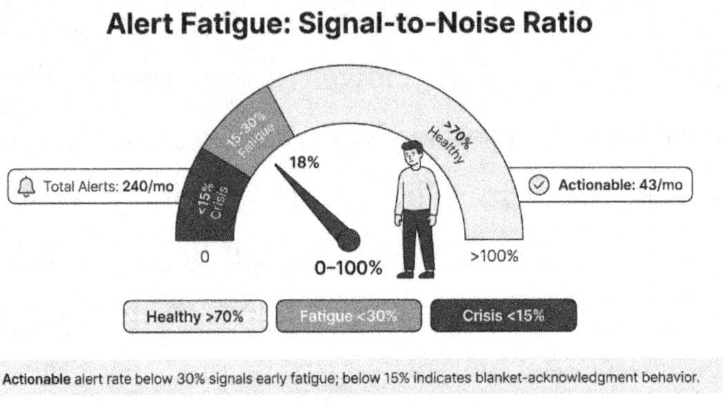

On-call dashboard at 2:47 AM: recommendation hit-rate SLO breach, index staleness root-cause trace, and kill-switch trigger sequence

11.2 Why It Matters for Production Systems

AI incidents are not classic SaaS incidents. A web service going down is immediately visible: requests return 5xx, latency spikes, dashboards turn red. AI failures differ in almost every dimension that matters for incident response.

The most dangerous difference is silent failure. A recommendation model whose quality has degraded by 30 percent continues to return HTTP 200 responses at normal latency. A language model that has begun hallucinating citations still produces fluent text. A fraud detection model operating on stale features still makes decisions at the correct p99. Detecting quality degradation requires measuring the quality of outputs themselves, not the health of the machinery producing them. Ground-truth lag compounds this: in production AI, ground truth often arrives hours, days, or weeks after inference. A credit model scores an application; the outcome arrives eighteen months later. This lag creates a window in which the model may be systematically wrong and no alert fires. The engineering response is to instrument leading indicators: input feature distribution shifts measured by PSI or KS-statistic, output score distribution changes, and evaluation scores from an automated eval harness against a golden dataset.

Severity is also ambiguous. When a web server returns 500 errors, the severity maps directly to the error rate. For AI systems, severity is a function of business context. A recommendation hit-rate drop at 3 AM during low traffic may be SEV-3. The same drop during a product launch at noon may be SEV-1. Severity cannot be defined by threshold value alone -- it requires a matrix that encodes business context, traffic volume, and regulatory exposure alongside the metric value. And unlike traditional outages, AI failures rarely generate immediate user complaints: a 15 percent recommendation degradation produces a slow erosion of click-through rate that only surfaces in the weekly analytics review. By this time, the blast radius has extended across millions of impressions. Building the discipline to detect and respond faster requires purpose-built SLIs, error-budget tracking, tested kill switches, and continuous improvement through blameless retrospectives.

The signal space is also far wider than traditional SRE practice. Classic golden signals—latency, error rate, saturation, and traffic— are necessary but not sufficient. AI observability adds input drift across dozens of features, output distribution shifts, eval score regressions, refusal rates, jailbreak detection rates, token cost per request, and retrieval precision in RAG pipelines. Without deliberate discipline, this expanded space produces an alert catalog that overwhelms any on-call rotation. Signal selection, threshold calibration, and alert lifecycle management are ongoing responsibilities of the platform.

11.3 Alert Design for AI Systems: SLI Selection, Threshold Strategies, and Burn-Rate Alerting

Every alert in your catalog should exist because it defends a specific SLO. If your team cannot name the SLO an alert defends, retire the alert. That discipline separates an alert catalog that helps from one that buries the on-call rotation in noise.

Selecting SLIs Across Five Dimensions

AI SLI selection must cover at least five dimensions. Latency SLIs measure inference time at the p50, p95, and p99 percentiles, shown as histograms rather than averages — averages hide the long tail that breaks SLAs for users who encounter it most. Drift SLIs measure the distance between the current input distribution and the reference distribution from training, using the Population Stability Index (PSI) or the Kolmogorov-Smirnov statistic. A PSI above 0.2 on a critical feature is a standard warning threshold, and a PSI above 0.25 typically triggers a critical alert. Eval score SLIs measure the output of an automated evaluation harness running on a cadence against a golden dataset; instrument them with OpenTelemetry custom metrics so they appear in unified dashboards alongside latency and error rate. Refusal and jailbreak SLIs apply to language models in sensitive contexts — both require a response classifier that needs versioning and validation, like any other model in the stack.

Cost SLIs track token consumption and per-request inference cost; context-window bloat bugs can multiply your inference bill by 10x within hours.

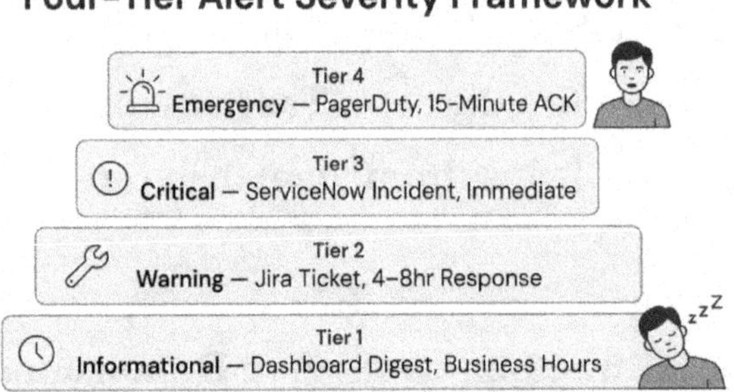

Tier assignment is a governance policy: only real-time business harm justifies an emergency page.

SLI taxonomy for AI systems: latency percentiles, drift statistics, eval scores, refusal and jailbreak rates, per-request cost

Threshold Strategies and Multi-Burn-Rate Alerting

Production AI alert design needs two threshold strategies working in layers. Deterministic thresholds -- a fixed value that fires when the metric crosses it -- are transparent, auditable, and easy to reason about under time pressure, but static. A model serving seasonal traffic will have different baseline input distributions in November than in July; a fixed drift threshold calibrated for July will result in false positives throughout the holiday season. Anomaly-detection thresholds learn from recent history and automatically adapt to changing baselines, but are harder to audit in a post-incident review. Layer them: use deterministic thresholds for hard floors that must never be crossed, regardless of baseline (a jailbreak rate above 5 percent is an emergency regardless of historical patterns), and use anomaly detection for quality metrics that drift with data seasonality.

The most powerful SLO-based pattern is multi-window, multi-burn-rate alerting: alert when the rate of error-budget consumption indicates you will exhaust the budget before the end of the SLO window. If your hit-rate SLO is 95 percent availability above 0.52 over a 30-day window, your error budget is roughly 36 hours. A burn rate of 14.4 exhausts it in approximately 50 hours. Implement two alert rules: a fast-burn rule that fires when the burn rate in the last one hour AND the last five minutes both exceed 14.4x; and a slow-burn rule that fires when the burn rate in the last six hours AND the last 30 minutes both exceed 6x—requiring both windows to agree on filters transient blips.

ITSM Integration by Alert Severity

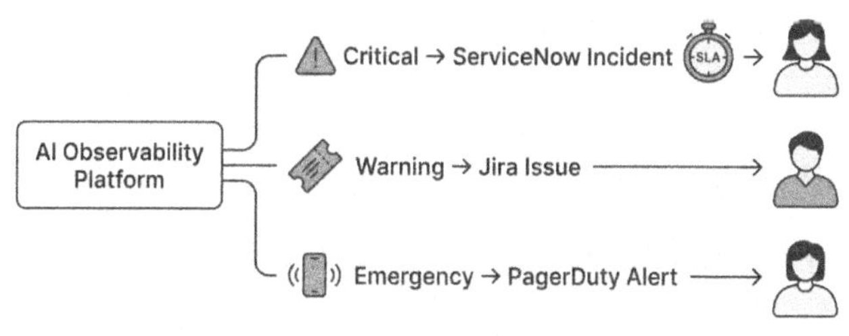

AI observability alerts must route through existing ITSM workflows, not parallel ad-hoc notification channels.

Multi-window burn-rate alert architecture: fast-burn and slow-burn rules layered over a 30-day error budget, dual-window agreement requirement.

Alert Hygiene to Prevent Pager Fatigue

Alert fatigue sets in when genuinely critical pages compete for attention against dozens of routine nudges. Treat the alert catalog as a managed artifact with a quarterly review cadence: every alert needs a named owner, a mapped SLO it defends, a documented notification channel and severity level, and a last-reviewed date. Measure the actionable alert rate weekly: the fraction of pages that resulted in a human taking a documented action. Below 50 percent, the catalog has a noise problem; below 30 percent, on-call engineers have likely entered blanket-acknowledgment behavior—acknowledging to silence the page without investigation. Route alerts using a four-tier model: informational to a Slack channel; warning-tier to a Jira ticket; critical-tier to a page with a 15-minute acknowledgment SLA; emergency-tier to both primary and secondary on-call with auto-escalation to the incident commander at 20 minutes.

11.4 Incident Response Playbooks for AI: Severity Matrix, Kill Switches, and Comms Protocol

A playbook is a pre-written, step-by-step response procedure for a specific class of failure. Without playbooks, every responder improvises under time pressure, producing longer MTTR and incomplete documentation. For regulated

AI systems, the playbook is also a compliance artifact: the absence of a documented, followed response procedure is itself a finding.

The AI Severity Matrix

The severity matrix must encode three variables: the metric value, the business context of the affected model, and the traffic volume or time window during which the failure occurs. A threshold-to-severity mapping alone is insufficient.

SEV-4 covers informational conditions: a metric trending unfavorably but still within SLO bounds. No page is generated. SEV-3 covers warning conditions: a metric approaching SLO threshold at a burn rate that will exhaust the error budget within seven days. SEV-3 creates a tracked ticket and notifies the model owner during business hours. SEV-2 covers conditions with confirmed or highly probable user impact: an SLO threshold crossed, a kill switch eligible for triggering, or a data pipeline failure affecting feature freshness. SEV-2 pages the primary on-call with a 15-minute acknowledgment SLA and automatically creates an ITSM incident. SEV-1 covers conditions that cause real-time business harm: a regulated model producing outputs outside documented bounds, a spike in jailbreak rate, or a cost SLO breach that threatens to exhaust the daily token budget within the hour. SEV-1 pages both primary and secondary on-call simultaneously, auto-notifies the incident commander, and triggers an attestation freeze on the affected model and its dependent pipelines. Encode business context as a severity multiplier via metadata tags on each production model: business-criticality tier, traffic-sensitivity window, and regulatory classification. A 10 percent drop in hit rate at 4 AM during off-peak traffic may be SEV-3. The same drop during peak transaction hours for a payment-adjacent model is SEV-1.

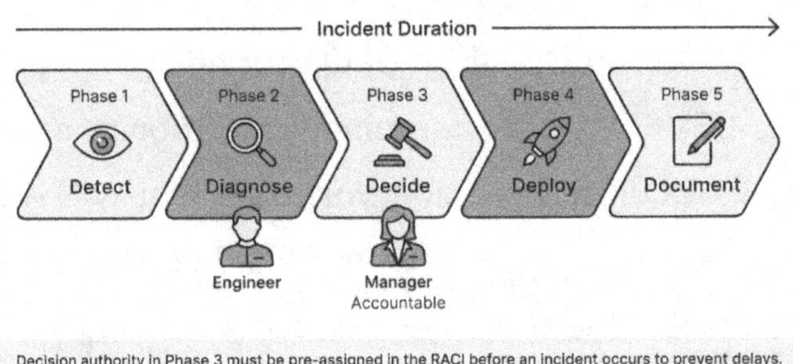

AI severity matrix: metric thresholds adjusted by business-context multipliers -- criticality tier, traffic window, regulatory classification

Kill-Switch Design and Fallback Model Routing

A kill switch is a pre-built, tested mechanism that immediately stops a model's production traffic and routes it elsewhere. It does not fix the problem, but it contains the blast radius while the fix is prepared. Every production model serving real-time traffic should have a kill switch before it goes live -- retrofitting one during an active incident is one of the most common root causes of extended MTTR. The routing target defines where traffic goes: a fallback model (a simpler, pre-validated model that trades quality for reliability), a deterministic fallback (a rule-based system that makes conservative decisions without a model), or a blue/green swap to a previously stable version. For language models, the deterministic fallback matters especially because degraded coherence is invisible to latency monitoring. The trigger must be a single command or button click -- no multi-step approval chain, no deployment pipeline run. Test it in staging monthly; an untested kill switch is a theoretical control. The recovery path specifies who repairs the primary model and what validation gate (typically a canary deployment at 5 percent of traffic for 30 minutes) must pass before full cutover.

AI Incident RACI Matrix

	ML Platform Eng	Model Owner	AI SRE Lead	Business Stakeholder
Technical Investigation	R	A	C	I
Remediation Decision	R	A	C	I
Business Impact	R	A	C	I
Regulatory Notification	R	Must Assign	I	I

Regulatory notification accountability must be assigned before an incident, not discovered during one.

Kill-switch architecture: primary model, fallback model tier, deterministic fallback tier, traffic routing layer, single-command trigger

Rollback Mechanics, Customer-Comms Protocol, and Attestation Freeze

Rolling back to a prior model version requires specific commands in the runbook, not vague directives. Use canary deployments for all blue/green swaps: promote the target version to 5 to 10 percent of traffic, observe SLIs for 15 to 30 minutes, then promote to full traffic. The staged approach prevents the scenario where the rollback target itself is problematic. Customer-comms protocol must be pre-defined for any SEV-1 or SEV-2 incident: who drafts the message (model owner or product on-call, not the technical on-call engineer), what approval is required, which channel it goes to, and the cadence (initial notification within 30 minutes of SEV-1 declaration, updates every 30 minutes until resolved). An attestation freeze is a directive issued when a model may have produced outputs requiring regulatory documentation; when active, no deployment, no model promotion, and no configuration change is permitted on the affected system or its upstream dependencies until the incident commander or compliance team lifts it.

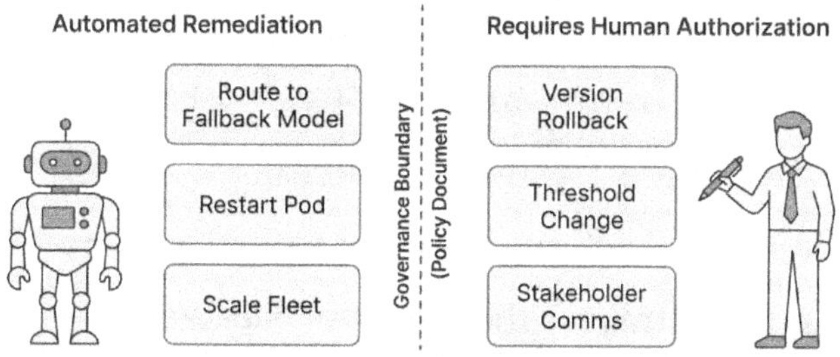

Automated actions are **reversible** infrastructure fixes; any change to **model logic** requires documented human authorization.

Rollback and canary promotion workflow: model registry version history, canary traffic split, SLI observation window, full-cutover gate

11.5 Post-Incident Review: Blameless Retros, Replay-From-Trace, and Eval Expansion

The post-incident review is where individual failures become organizational intelligence. Each structured review produces permanent improvements to runbooks, alert thresholds, eval coverage, and on-call rotation design.

Blameless Retros and Drift Root-Cause Analysis

A blameless retrospective operates on the principle that system failures are produced by design gaps, not by individual engineer failures. When Priya triggered the kill switch at 2:47 AM, the correct question is not "why didn't she catch the index staleness earlier?" but "why did the index staleness not surface in pre-deployment checks, and why did monitoring wait 18 minutes to fire?" The first question produces blame. The second produces system improvements.

Structure the retro around five sections: incident timeline, proximate cause, systemic causes, impact quantification, and action items. Reconstruct the timeline from telemetry -- traces, the metrics backend, deployment event logs, and alert firing history -- rather than from memory. Drift root-cause analysis traces a quality regression to one of four categories: input distribution shift (traffic differs from training data), label or reference shift (the ground-truth definition changed), infrastructure or serving artifact (a deployment altered model behavior), or model capability limitation. Each category has a different remediation path, and confusing input distribution shifts due to model capability limitations lead to the wrong fix. Investigation tools include PSI and KS-statistic comparisons, output score distribution comparison, and slice analysis.

Due: 5 Business Days

Post-Incident Review Document Structure

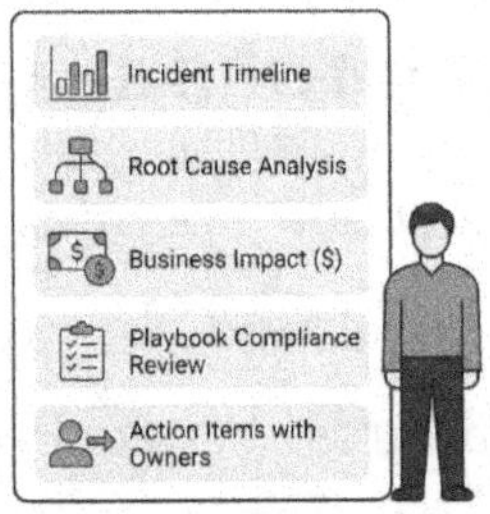

Each review section targets a specific program improvement: threshold, playbook, RACI, or pipeline change.

Drift root-cause decision tree: input shift, label shift, infrastructure artifact, model capability gap, with remediation paths.

Replay-From-Trace and Eval Expansion

Replay-from-trace is underused in most organizations. If your serving layer emits full trace records -- input features or prompt, model output, inference metadata, and timestamp -- you can replay the exact inputs from the incident window through any model version after the fact. Implement replay-from-trace as a pipeline component that reads from your trace store, filters by model ID and time window, and submits inputs to a target model version. If replaying the prior model version produces correct outputs, the regression is in the current version—confirming the root cause without guesswork. Apply PII masking before writing to the trace store and define retention policy in advance (30 to 90 days).

Eval expansion as remediation uses incidents to grow the golden dataset and the eval harness. Every incident that reveals a failure mode the eval harness missed is evidence of a coverage gap. After the blameless retro, add representative examples from the incident window to the golden dataset, and write eval cases that would have caught the regression. Verify that the expanded harness produces a failing result against the pre-incident model version and a passing result against the fixed version. Use incident timeline documentation as source material for runbook generation: after five or ten incidents of a given class, the resolution pattern can be formalized with specific commands, dashboard links, and acceptance criteria.

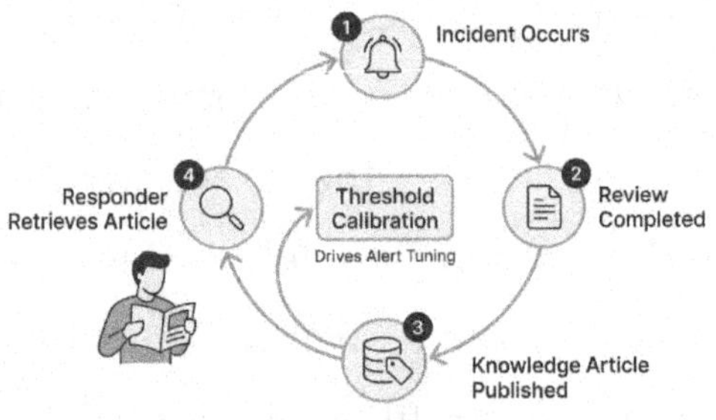

Replay-from-trace pipeline: trace store ingestion, input replay to candidate model, output comparison showing incident-window regression vs. fixed model

On-Call Rotation Design and MTTD/MTTR Tracking

Design AI on-call rotations around model domain teams: primary on-call from engineers familiar with the stack, secondary on-call from the ML platform team for infrastructure-level escalations, and an incident commander (paged only for SEV-1) with authority to trigger kill switches and order rollbacks. Rotate primary on-call weekly with a hand-off meeting covering active incidents, recent deployments still in their observation window, and upcoming events that may produce unusual traffic patterns.

Track action items from blameless retros in the main engineering backlog, reviewed in sprint planning. Track MTTD and MTTR per model domain and per root-cause category. An MTTD consistently above 15 minutes suggests that thresholds are too conservative or that a failure class has been uncovered. An MTTR consistently above 60 minutes for SEV-2 incidents suggests incomplete runbooks, untested kill switches, or insufficient on-call domain context.

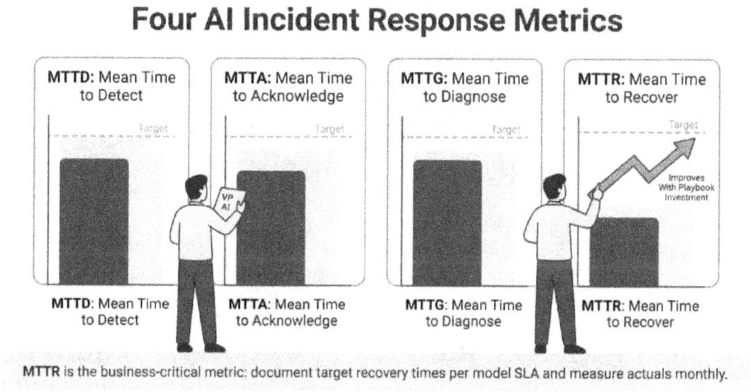

Four AI Incident Response Metrics

MTTR is the business-critical metric: document target recovery times per model SLA and measure actuals monthly.

On-call health dashboard: MTTD and MTTR by model domain, actionable alert rate trend, open post-incident action items by age.

11.6 Technical Team Checklist: AI On-Call Readiness

- **Action:** Define SLIs for every production model across latency (p50/p95/p99), input drift statistic, eval score, refusal rate, and per-request cost; register each with a named SLO threshold. **Owner:** ML Platform Eng. **Done**

when: All five SLI dimensions are queryable in a unified dashboard alongside infrastructure golden signals, with SLO thresholds annotated as reference lines.

- **Action:** Implement multi-window multi-burn-rate alert rules for all quality SLOs using a fast-burn rule (14.4x over 1-hour and 5-minute windows) and a slow-burn rule (6x over 6-hour and 30-minute windows), requiring both windows to agree before a page fires. **Owner:** ML Platform Eng. **Done when:** Rules are tested against historical data and confirmed to fire within 10 minutes of each recorded SEV-2 or higher incident.
- **Action:** Build and test a kill switch for every production model serving real-time traffic, wired to the feature flag or traffic-routing layer and triggerable with a single command, with no additional approval required for fallback or deterministic routing targets. **Owner:** ML Platform Eng. **Done when:** Kill switch is tested monthly in staging; on-call runbook includes the exact command and expected SLI recovery timeline.
- **Action:** Write runbooks for every alert type in imperative form with specific dashboard links, CLI commands, and acceptance criteria for each step, linked directly from the PagerDuty alert configuration. **Owner:** SRE / On-Call. **Done when:** Every PagerDuty alert has a linked runbook that a new on-call engineer with model domain context can execute without additional guidance.

- **Action:** Instrument the serving layer to capture full trace records with PII masking before write, configure a 30-to-90-day retention policy, and build the replay-from-trace pipeline. **Owner:** ML Platform Eng. **Done when:** Replay-from-trace pipeline is functional and was used in at least one post-incident review to confirm root cause.
- **Action:** Implement the four-tier severity matrix encoding business criticality tier, traffic sensitivity window, and regulatory classification as metadata tags read at alert-routing time. **Owner:** ML Engineer. **Done when:** Pipeline severity matches manually assigned severity for at least 90 percent of a sample of 50 historical incidents.
- **Action:** Configure PagerDuty escalation chains with primary on-call, secondary on-call, and incident commander tiers, with auto-escalation at 10 minutes to secondary and 20 minutes to incident commander for SEV-2 and above. **Owner:** SRE / On-Call. **Done when:** The escalation chain is tested with a synthetic page, and all three tiers receive notifications in the correct sequence.
- **Action:** Run blameless post-incident reviews within five business days for every SEV-2 and above incident; store reviews in a queryable knowledge base tagged with model ID, alert type, root cause category, and resolution action; track action items in the engineering backlog. **Owner:** SRE / On-Call. **Done when:** Review completion

rate exceeds 95 percent for SEV-2 and above incidents over a rolling 90-day window.

- **Action:** After every SEV-2 or higher quality regression, expand the golden dataset with representative examples from the incident window and update the eval harness so new cases produce a failing result against the pre-incident model version. **Owner:** ML Engineer. **Done when:** Eval coverage report shows no incidents in the last 90 days whose failure mode was not already represented in the golden dataset.
- **Action:** Review and calibrate the alert catalog quarterly, retiring or adjusting thresholds for any alert whose actionable rate fell below 50 percent, and publishing the updated catalog with documented rationale. **Owner:** ML Platform Eng. **Done when:** Alert catalog review is a standing quarterly agenda item with written decisions for every modified or retired alert.

11.7 Team Conversation Starters

- For each production model, can you name the SLO it is governed by, the current burn rate against that SLO, and the specific runbook step that executes if the fast-burn alert fires at 3 AM on a Saturday?
- If the on-call engineer triggers the kill switch for the recommendation ranker tonight, where does traffic go, what quality degradation should users expect, and who

is responsible for diagnosing and re-enabling the primary model?

- When was the last time you replayed a representative sample of production requests through staging before a promotion, and what would replay-from-trace have shown if it had run during the most recent SEV-2 incident?
- What is your actionable alert rate for the last 30 days, and which three alert rules account for the most noise? What change would most improve signal-to-noise without reducing coverage?
- How many post-incident action items from the last 90 days are still open, and is the root cause prioritization pressure, ownership ambiguity, or absence from sprint planning?
- If a regulated model fires a SEV-1 alert during business hours, who has the authority to declare an attestation freeze, and which systems does the freeze lock automatically?

11.8 Key Takeaway

The engineering challenge in AI incident response is not building more monitors; it is building fewer, better-calibrated ones and pairing each with a runbook, a tested kill switch, and a severity classification that encodes business context. Multi-window, multi-burn-rate alerting signals how quickly you are consuming your error budget before it's gone. Kill switches that route to deterministic fallbacks contain blast radius without heroics. Replay-from-trace turns incident timelines into ground truth for root-cause confirmation. Blameless retros that feed action items into the engineering backlog and examples into the eval harness compound investments, making the second incident of a given class cheaper than the first.

12 AI Model Governance, Compliance, and Audit Readiness

12.1 Opening Scenario

Priya is a security engineering lead at a mid-sized financial services platform. Her team runs twelve production models: credit underwriting, fraud scoring, transaction categorization, and a handful of LLM-backed customer-facing features. It is a Tuesday morning when the EU AI Act audit team books a three-day on-site. The lead auditor sends a pre-audit questionnaire with a single sentence: "Please provide evidence that every production prediction generated by high-risk systems over the preceding twelve months is reconstructible, including the model version, prompt template version if applicable, input features, and evaluation gate results at deploy time."

Priya does not panic. She opens her attestation pipeline dashboard.

Every model version her team ships is signed with a content-addressed artifact hash before it is allowed into the production registry. The signing key is held in a hardware security module. The signature event, the SHA-256 digest of the model, the eval gate run ID, and the signing key fingerprint are written together into an append-only log backed by an immutable object store. Every prompt template version for the LLM features is hashed, stored in the content-addressed template registry, and its version is pinned into every inference trace via an OpenTelemetry span attribute called prompt_template_sha. The eval harness runs a suite of golden-dataset evaluations on every candidate model; the pass/fail result, the eval dataset SHA, and the evaluator version are committed to the same append-only log as a separate event linked to the model version by a foreign key. Nothing in the log can be modified or deleted before the retention clock expires.

She exports the audit evidence package directly from the pipeline: a JSON-Lines archive, one line per signed event, covering every model deployment and every eval run for the year. The replay export includes a Merkle root that lets auditors verify log integrity without reading every record. She walks them through the pipeline architecture in thirty minutes. They spend the rest of the three days reviewing the records rather than arguing about whether records exist.

That is what audit-ready governance engineering looks like: not a binder assembled under deadline pressure, but artifact pipelines that produce cryptographically verifiable evidence as a routine byproduct of shipping software.

AI Regulatory Framework Coverage Map

	Performance Monitoring	Drift Records	Decision Audit Log	Access Controls	Lineage
SR 11-7	●	●	●	●	●
NIST AI RMF	●	○	●	○	○
EU AI Act	●	●	●	●	●
HIPAA	○	○	●	●	○
OMB M-26-04	●	●	●	●	●

● = Required
○ = Recommended

Every major AI regulatory framework requires structured observability records as primary compliance evidence.

Attestation pipeline: model version signing, eval gate linkage, and append-only log write at deploy time

12.2 Why It Matters for Production Systems

The governance failure mode that destroys engineering teams is rarely the dramatic one. It is the quiet failure: a team has been running a model for fourteen months; the model has drifted, and when the regulator asks for records, there are none. The team was monitoring. They were not retaining, signing, or structuring the evidence in a form anyone outside the team can reason about.

Production AI systems generate telemetry continuously. Most teams route it into dashboards and short-retention metrics pipelines optimized for operational alerting. That is correct for on-call response. It is the wrong architecture for governance, because dashboards are not records. A time-series database with a thirty-day retention window is not audit evidence. A screenshot of a Grafana panel attached to a Jira ticket is not a reconstruction of what a model did on a specific date for a specific input.

Regulatory frameworks are converging on a concrete requirement: the ability to reconstruct the context of any automated decision. EU AI Act Article 12 requires high-risk AI systems to automatically log their outputs with metadata sufficient to identify the system version, input data characteristics, and decision context. NIST AI RMF Measure 2.5 requires that measurement results be documented with sufficient specificity to support external review. HIPAA Security Rule section 164.312(b) requires audit controls that record and examine activity in systems that contain electronic protected health information. None of these are satisfiable with a time-series dashboard. All require structured, retained, tamper-evident records.

The engineering consequence of treating governance as a post-hoc concern is that the team builds a remediation system under time pressure, usually after an audit finding. Remediation is more expensive than prevention, just as adding observability to a system after it fails is more expensive than instrumenting it at the start.

There is a subtler failure mode: a team retains records, but the records are not linked. The model registry records the model version. The eval harness has a record of the eval run. The inference log records predictions. None shares a common key. When an auditor asks, "Show me the eval results for the version that produced this decision," the answer requires manually correlating three separate systems. Manual correlation is not reproducible, is not auditable, and breaks when the person who knows the mapping leaves the team. The fix is a shared artifact identifier, a model version ID, or a content hash that propagates through every system that touches the model from the registry to the eval harness to the inference trace to the audit log.

Governance engineering is ultimately a lineage problem. Every artifact in the AI delivery pipeline has a provenance chain: training data version, feature transformation code version, model training run ID, trained artifact hash, eval gate results, deploy gate approval, model server config version, and prompt template version if applicable. The governance requirement is that this chain be navigable in both directions. Forward navigation answers "what decisions did this model version produce?" Backward navigation answers "what data and code produced the model that made this decision?" The SBOM pattern from the software supply chain world applies directly: the AI artifact SBOM records every versioned input that produced a given model, and it is signed and retained alongside the model artifact itself. Both directions need to be queryable from a single system of record with immutable records.

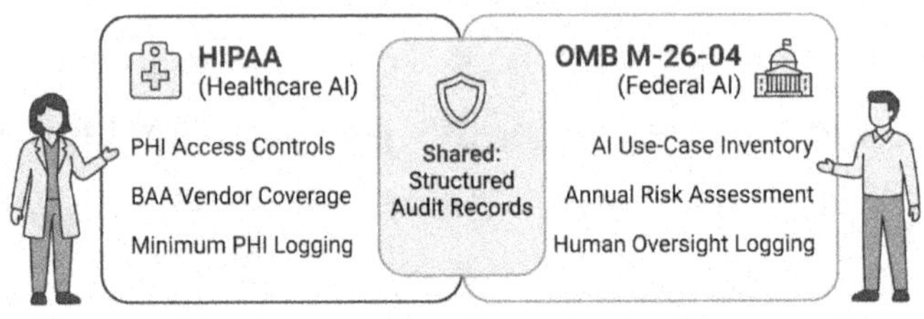

Both frameworks require structured queryable monitoring records accessible on demand during examinations.

Lineage chain from training dataset version through feature transform, model artifact, eval gate, and inference trace to audit log

12.3 Audit-Log Architecture: Immutable Stores, Hash Chains, and Retention

An audit log for an AI system is not the same as the application log. Application logs support engineers' debugging live systems. Audit logs support reviewers in reconstructing past events. Getting this separation wrong is one of the most common governance engineering mistakes: teams configure their existing log aggregation pipeline as the audit log, then discover the retention policy is thirty days, the logs are mutable by anyone with the right IAM role, and there is no schema enforcement. None of that is fixable without rearchitecting the pipeline.

Content Addressing and Hash Chains

Content addressing is the foundation of a tamper-evident artifact store. Instead of naming an artifact by a mutable identifier, the store derives the identifier from the content itself using SHA-256. The artifact identifier is the hash. If the artifact changes, the hash changes. A reference to a content-addressed artifact is also a verification commitment: you can confirm that what you retrieved matches the reference, without trusting the storage system's integrity. Apply content addressing to every artifact in the pipeline: training dataset snapshot, feature transformation code at VCS commit, trained model after serialization, eval dataset before harness consumption, and prompt template after normalization. Each hash becomes an event payload field in the audit log.

A hash chain links successive log entries so that tampering with any entry invalidates all subsequent entries. Include the hash of the previous log entry as a field in each new entry. A verifier replays the chain from genesis to the current head and confirms no entry was modified or deleted. Merkle trees allow proof of inclusion for a specific entry without replaying the entire chain. Implement hash chaining at the audit log writer level: the service receives events, assigns a sequence number, hashes the full event payload, appends the previous entry hash, signs the record with a KMS-backed key, and writes the signed record to the immutable store. The signing key fingerprint is included in every record, so key rotation does not break verification of historical records.

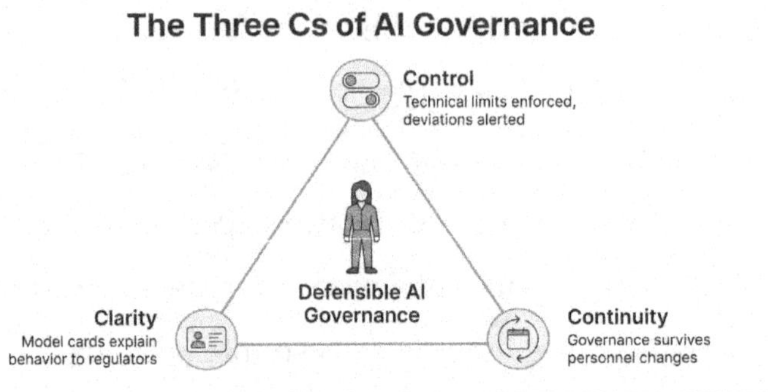

Control, **Clarity**, and **Continuity** transform compliance from a reactive task into sustained operational discipline.

Hash chain structure: each audit log entry includes the SHA-256 of the prior entry, event payload hash, and HSM signature

Signing, Key Custody, Retention, and Replay Export

Sign audit log entries with an asymmetric key held in a cloud KMS or on-premises HSM. The signing service calls the KMS sign API, which never exposes the raw private key. The public verification key is published in a well-known location and included in the genesis record. Define key custody explicitly: which team has access, who can rotate it, and the rotation cadence. Key rotation events are themselves audit log entries that record the old and new fingerprints, the timestamp, and the operator identity. For multi-tenant platforms, hierarchical signing allows a root key to sign team-level intermediate keys, enabling per-team rotation without platform-wide coordination.

Retention policy is a first-class engineering requirement. EU AI Act high-risk systems require up to 10 years in some implementations; HIPAA requires 6 years; FedRAMP High requires 3 years online and 5 years total. Build the retention policy as a configuration value in the audit log service rather than a storage bucket lifecycle rule. The service enforces its own minimum retention period and rejects delete requests within the window. Use a schema registry with explicit version fields in every record.

Replay export is the mechanism for producing evidence packages. The audit log service exposes an export API that accepts a time range, a set of model identifiers, and an optional event-type filter, returning a deterministic byte sequence of matching records, along with a Merkle root. The export format should be auditor-friendly: JSON-Lines with a companion verification script that the auditor can run to confirm the Merkle root matches the records.

Five Categories of AI Audit Evidence

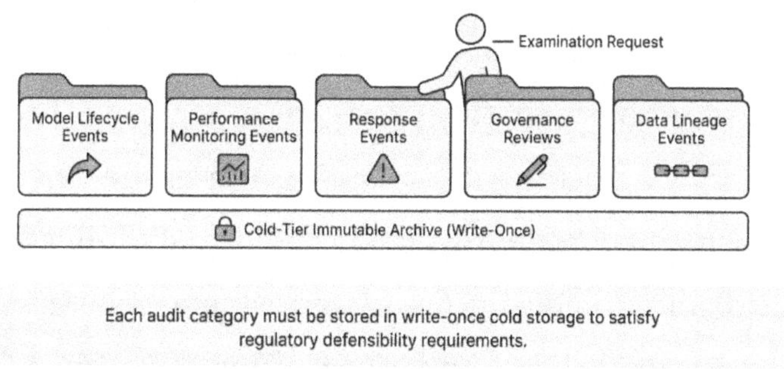

Each audit category must be stored in write-once cold storage to satisfy regulatory defensibility requirements.

Replay export pipeline: time-range query against append-only log, Merkle root computation, and signed evidence package output

12.4 Programmatic Policy Enforcement: Policy-as-Code, Attestation Gates, and Data-Residency Controls

Governance policy expressed only in a document is not governance. It is a statement of intent that may or may not be followed depending on whether the engineer reads it correctly under deploy-time pressure. Programmatic policy enforcement evaluates the policy as code at the exact moment a consequential action is attempted, blocks the action if the policy is violated, and records the evaluation result in the audit log regardless of outcome.

Policy-as-Code with OPA and Rego

Open Policy Agent (OPA) evaluates policies written in Rego against structured input data and returns an allow or deny decision, along with a structured reason. A pre-deploy attestation gate runs in the CI/CD pipeline before a model artifact is promoted to the production registry. The gate input is a JSON document containing the model metadata: artifact hash, eval gate result IDs and pass/fail statuses, model card completeness score, data residency classification, PII handling attestation, and signing key fingerprint. The

Rego policy checks each field. If the model card is missing a required field, the gate denies promotion. If the eval gate did not run against the current golden dataset version, the gate denies promotion. If the training data residency classification does not match the permitted deployment region, the gate denies promotion. Store the policy file alongside the model registry configuration, require governance engineer approval on any modifying pull request, and include the policy file hash in every attestation gate evaluation record.

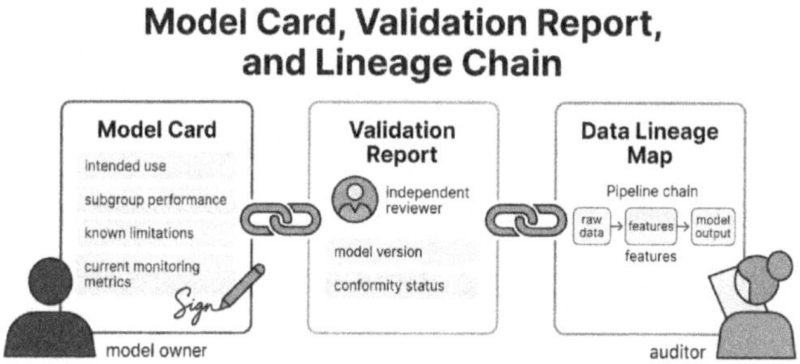

Model Card, Validation Report, and Lineage Chain

Model card, validation report, and lineage together form the evidence chain for any adverse-action regulatory inquiry.

OPA/Rego attestation gate in the CI/CD pipeline: policy evaluation against model metadata, allow/deny decision, and audit log write

Eval Gates, Deploy Gates, and PII Handling Policy

An eval gate enforces quality by running the candidate model against a pinned golden dataset and comparing results against policy-defined thresholds. The gate result is a structured record linking the model artifact hash, the eval dataset hash, the evaluator version, the metric results, the threshold policy version, and the gate decision. A future auditor asking "what were the bias metrics for the model version deployed on this date?" retrieves the answer from the audit log without re-running the evaluation. Chain the eval gate and deploy gate results into the artifact lineage record for the model version, so the lineage record is queryable as a single document: training data hash, model artifact hash, eval gate run ID and result, deploy gate run ID and result, signing key fingerprint, and production registry promotion timestamp.

A deploy gate runs immediately before the model server routes traffic to the new version, checking canary rollout SLO results, feature schema match against the model card, resource allocation verification, and required approval signatures.

PII handling policy in code means the inference pipeline cannot log a raw input containing a recognized PII entity without first passing it through a redaction filter. The classifier runs as a pre-logging filter in the inference span processor. Recognized entity types are replaced with a structured redaction marker that records the entity type and the character offset, but not the value. Data-residency policy enforcement prevents a model serving EU resident data from writing inference logs to a US-region object store. Implement this as a property of the inference trace writer:

the writer selects the output destination based on a residency label propagated through the OpenTelemetry context as a baggage item, set at the API gateway based on request origin or tenant configuration.

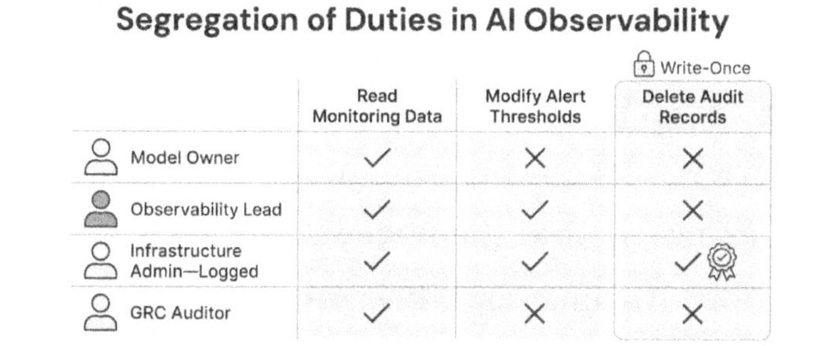

Segregation of Duties in AI Observability

	Read Monitoring Data	Modify Alert Thresholds	Delete Audit Records (Write-Once)
Model Owner	✓	✗	✗
Observability Lead	✓	✓	✗
Infrastructure Admin—Logged	✓	✓	✓
GRC Auditor	✓	✗	✗

Engineers who build models must not have unilateral control over the monitoring records used to assess them.

PII redaction pipeline: pre-logging classifier, span attribute replacement, and egress redaction policy applied at export time

12.5 Mapping to Regulatory Frameworks: EU AI Act, NIST AI RMF, ISO 42001, and Compliance Overlays

Regulatory framework mapping is an engineering task before it is a compliance task. The practical question is not "does our program comply with the EU AI Act" but "which controls does each framework require, what evidence satisfies each control, and is that evidence being produced, retained, and made queryable." The mapping links each requirement to a specific audit log event type, schema field, or pipeline gate. It is a versioned document maintained alongside the governance engineering runbook.

EU AI Act: Control-Implementation Mapping

EU AI Act Annex IV specifies technical documentation requirements for high-risk systems. Article 9 maps to the risk assessment event: a structured record from each quarterly review that documents the current risk posture, the metrics reviewed, and the reviewer's attestation. Article 10 maps to the training data lineage record: the content-addressed dataset hash, the data quality report run ID, and the data card. Article 11 maps to the versioned model card in the model registry. Article 12 maps to the inference trace span, enriched with the model version and prompt template version attributes. The conformity assessment requirement means the deploy gate record serves as the primary evidence artifact: a structured attestation that the pre-deploy checklist was completed, that documentation was present, and that an authorized reviewer signed the approval, with the reviewer's identity and timestamp retained for the full statutory period.

Observability-to-GRC Evidence Flow

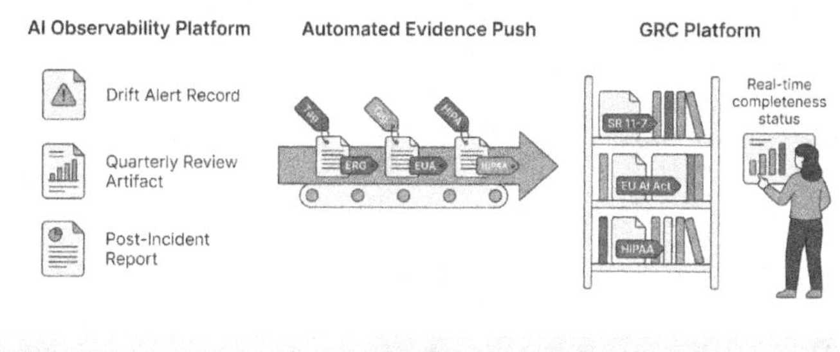

Automatic evidence tagging eliminates weeks of manual gathering before regulatory examinations begin.

EU AI Act Article mapping: regulatory requirement, control type, audit log event type, and evidence field for Articles 9 through 12

NIST AI RMF, ISO 42001, and Compliance Overlays

NIST AI RMF Measure 2.5 requires measurement results to be documented with sufficient specificity to support external review. Operationalize this by writing every eval harness run to the audit log with full provenance: eval dataset hash, evaluator version, metric definitions, numeric results, and comparison against the baseline. Measure 2.7 requires system performance to be evaluated on a defined cadence. Operationalize this as a scheduled pipeline job: it pulls model performance metrics, computes drift against the baseline window, queries the alert history, and publishes a structured review package to the model registry, with the model owner's sign-off recorded as a separate audit log event. Encode the NIST Governance accountability structure into the deploy gate Rego policy: production deployments require a sign-off event from a governance engineer role managed in the identity provider, so role changes are auditable. ISO 42001's Check function maps to the same quarterly review pipeline. Build the evidence export API to accept a framework parameter, labeling exported records with the relevant control identifiers, so the same evidence satisfies multiple framework audits.

SOC 2 Type II requires controls to operate effectively over the audit period. Build the audit log query interface to support population export: given an audit period, return all governance review events, deploy gate events, PII redaction events, and alert response events so the auditor can sample from a verifiable population. HIPAA's human oversight requirement means fully automated decisions are not permitted for certain clinical functions. Operationalize this as a classification gate that denies production promotion for fully automated HIPAA-covered models until the model card records a human review protocol. FedRAMP High requires continuous monitoring with monthly reporting; implement this as a scheduled pipeline job that generates a performance monitoring report and a POA&M update for each authorized model.

The "human oversight" control appears across all frameworks. The most common mistake is treating it as policy without a technical enforcement mechanism. Implement it as a staleness check in the deploy gate: if the most recent governance review event is older than the policy-defined maximum interval, the gate rejects any deployment, including rollbacks, until a fresh review event is recorded. The cadence is platform-enforced, not calendar-enforced.

Examiner Questions Mapped to Audit Evidence

Examiner Question		Required Evidence
Who monitors this model?	→	RACI Document
When last reviewed?	→	Quarterly Review
Show Q3 report.	→	Performance Report
Demographic parity proof.	→	Demographic Stats
Training data lineage.	→	OpenLineage Record

Mock examinations reveal evidence gaps before real regulators arrive; five standard questions map to five evidence types.

Human oversight control: staleness check in the deploy gate, review event schema, and governance cadence enforcement pipeline

12.6 Technical Team Checklist: Governance and Audit Engineering

- **Action:** Implement a content-addressed artifact store for all AI pipeline artifacts using SHA-256 hashing at write time, with the hash as the primary artifact identifier in the model registry. **Owner:** ML Platform Eng. **Done when:** Any artifact can be retrieved and re-hashed to confirm that the bytes match the registered identifier, and a CI check fails the build if the hash is missing or mismatched.

- **Action:** Deploy an append-only audit log service backed by an object store with write-once retention locks (S3 Object Lock COMPLIANCE mode or equivalent), with an HSM-backed signing key and a hash chain linking each entry to its predecessor. **Owner:** Security Eng. **Done when:** A tamper-detection test modifying a historical record produces a chain verification failure within a single verification run.
- **Action:** Write OPA/Rego pre-deploy attestation gates for all high-risk model classes, checking model card completeness, eval gate pass status, PII handling attestation, and data residency classification before promoting any artifact to the production registry. **Owner:** Governance Eng. **Done when:** A policy-failing artifact is rejected with a structured deny reason logged to the audit system, and a compliant artifact is promoted with the gate evaluation record in the audit log.
- **Action:** Instrument the inference pipeline to propagate model version ID and prompt template SHA as OpenTelemetry span attributes on every prediction span, and configure the span exporter to write these attributes to the audit log for each batch of inference records. **Owner:** ML Platform Eng. **Done when:** A query against the audit log for any model version ID returns the

inference volume and a sample of reconstructible prediction contexts from the past ninety days.

- **Action:** Implement PII classification and redaction as a pre-logging filter in the inference span processor, with entity-type and offset markers replacing raw PII values, and configure the egress redaction service to apply the appropriate policy per export destination. **Owner:** Security Eng. **Done when:** A red-team test injecting synthetic PII produces zero raw PII values in the exported audit log, with entity-type markers at the correct offsets.
- **Action:** Build a control-to-evidence mapping document linking each applicable regulatory requirement to a specific audit log event type, a specific schema field, and the required retention period, maintained in the governance runbook with annual governance and legal review. **Owner:** Governance Eng. **Done when:** Any member of the security engineering team can identify the audit log query required to satisfy a given regulatory control without asking the platform team.
- **Action:** Configure a staleness enforcement gate that rejects any deployment of a model whose most recent governance review event is older than the policy-defined maximum interval, logging the rejection with the model version ID and staleness duration. **Owner:** ML Platform Eng, Governance Eng. **Done when:** A test deployment of

a model with a deliberately stale review event is rejected with a structured error appearing in the audit log.

- **Action:** Implement an evidence export API accepting a regulatory framework parameter, time range, and model identifier, returning audit log records labeled with relevant control identifiers plus a Merkle root and companion verification script reference. **Owner:** ML Platform Eng. **Done when:** A mock audit exercise produces a self-verifiable package that an external reviewer with no production access can validate using only the exported records and the published verification script.

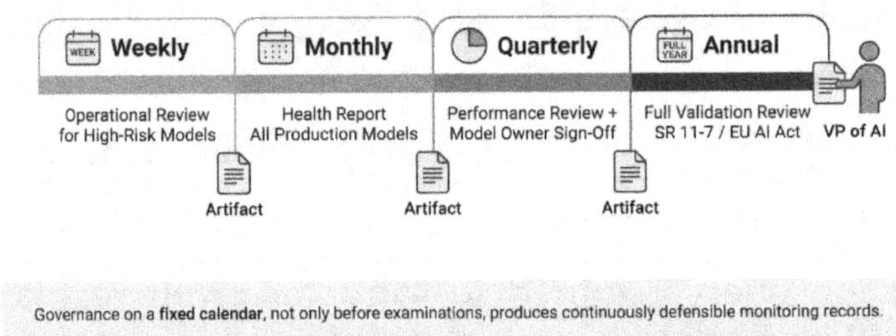

Governance engineering system map: artifact store, audit log service, OPA gates, evidence export API, and regulatory control mapping

12.7 Team Conversation Starters

- If a regulator asked us today to produce all inference records for model version X over the past twelve months, what exactly would we retrieve, how long would it take, and what format would the records be in?

- Which artifact in our current pipeline lacks a content-addressed identifier, and what class of audit question becomes unanswerable because of that gap?
- Our eval gate checks model quality, but does it record the eval dataset version and the evaluator implementation version in the audit log in a form that is queryable three years from now?
- If a governance engineer who knows nothing about our internal systems ran the evidence export API against our audit log, would they get a self-contained, verifiable package, or would they need to ask us to interpret it?
- Where in our inference pipeline could a PII value enter the span attributes without being caught by the pre-logging filter, and what is the blast radius of that gap for our HIPAA compliance posture?
- Our deploy gate enforces policy at promotion time, but does the staleness check actually block a rollback to a model version whose governance review has expired, or does the rollback path bypass the gate?

12.8 Key Takeaway

Governance engineering is artifact pipeline engineering. Every requirement in the EU AI Act, NIST AI RMF, ISO 42001, HIPAA, SOC 2, and FedRAMP resolves to the same set of problems: produce a content-addressed artifact, write a signed event to an append-only log, enforce a policy gate before a consequential action, retain the record for the required period, and make it queryable in the form an external reviewer needs. Teams that wire these mechanisms into the delivery pipeline at the start treat compliance evidence as a byproduct of shipping software. Teams that bolt them on after an audit finding pay the full retrofit cost plus the remediation cost. The attestation pipeline, the hash chain, the OPA gate, and the evidence export API are not governance overhead. They are the production-grade infrastructure that enables operating AI systems at scale without losing the ability to account for what those systems did.

13 AI Cost Optimization and FinOps for Enterprise AI Workloads

13.1 Opening Scenario

It is a routine Thursday when Priya, the SRE lead for the ML platform team, opens the weekly infrastructure digest and spots a number that stops her cold: cloud spend on a single LLM inference endpoint has tripled over the past 14 days. The alerting pipeline never fired because the absolute spend threshold was anchored to last month's baseline, and no one updated it after the most recent feature launch.

She pulls the endpoint's telemetry. Request volume is flat. Latency is flat. Error rate is clean. But the token-count histogram tells a different story: tokens-in-per-request has ballooned by roughly 400 percent. The median prompt length two weeks ago was 280 tokens. Today it sits at 1,100.

Priya pivots to the per-tenant breakdown. That view does not exist. Token counts are aggregated at the endpoint level with no slice by tenant identifier. She implements a quick change: a span attribute on every LLM gateway call that propagates the tenant ID through to the token-count metric. Within an hour, she has a per-tenant cost-per-request dashboard.

One tenant accounts for 83 percent of the token growth. Tracing their request payloads reveals the root cause: a recently shipped integration triggers an embedding call on every keystroke in a text field, passing the full document as context rather than just the incremental delta. No caching layer exists. Every keystroke is a fresh, full-context round-trip to the model.

The fix ships the same afternoon. A semantic cache keyed on a normalized-prompt hash deflects near-duplicate requests without hitting the model. A per-tenant rate limit at the gateway caps burst-embedding requests to 20 per second, with backpressure signaling the client to debounce. Within 48 hours, the spend drops by 70 percent and then holds. The operational SLIs never blinked. This is the normal failure mode when cost is not treated as a first-class runtime signal.

The AI Infrastructure Cost Iceberg

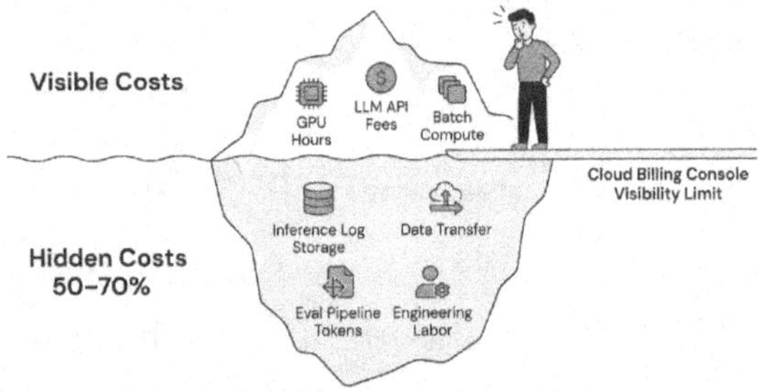

AI FinOps programs tracking only the cloud bill miss 50–70% of actual total AI cost of ownership.

Per-tenant token-count telemetry identifying a single integration responsible for 83% of endpoint cost growth

13.2 Why It Matters for Production Systems

AI cost is a runtime signal, not a billing-team report. Finding a cost anomaly 48 hours after it starts is an engineering problem you can instrument and fix in a sprint. Finding it six weeks later on a cloud invoice is a budget conversation that happened too late to prevent the waste. Traditional application infrastructure has predictable unit economics. AI inference breaks that model in four directions at once.

Token non-linearity is the first breakage. Two requests to the same LLM endpoint can incur cost ratios of 10:1 or higher, depending on prompt and response lengths and whether the KV cache was warm. A single request that accidentally includes a 20,000-token context block costs as much as 40 normal requests. Without per-request token instrumentation, the only detection path is to notice an incorrect monthly bill.

Shared infrastructure opacity is the second breakage. A GPU inference cluster serving 12 model teams has no natural cost-attribution boundary. When GPU utilization climbs from 45 percent to 78 percent, you know something changed, but not which team drove it, unless every inference request carries a feature-level tag preserved through the metric pipeline. Without that tag, right-sizing decisions are guesses, and chargebacks are impossible.

Agent-loop runaway is the third breakage. An autonomous agent calling tools in a loop can exhaust a monthly token budget within hours if a termination condition is missing or a tool returns a malformed response, trapping the agent in a retry cycle. The incident playbook for runaway agents does not exist at most organizations because the problem has not been treated as an engineering problem with a kill switch.

Cache invisibility is the fourth breakage. Semantic caching can deflect 30 to 60 percent of LLM calls at near-zero cost, but those savings are invisible without a cache-hit-rate metric wired into the same dashboard as cost-per-request. If the hit rate drifts from 40 percent to 12 percent because a prompt template changed, spend climbs silently. Cache hit rate must be a first-class SLI alongside latency and error rate.

None of these failure modes is detectable through standard cloud billing dashboards. Cloud bills do not report cost-per-prediction, cost-per-tenant, tokens-per-feature, or cache hit rate. Getting from a cloud bill to an actionable engineering diagnosis requires the same telemetry pipeline the platform team already runs for latency and availability, extended to carry cost-relevant attributes. Cost signals belong in the observability stack alongside latency signals. When a cost-per-request metric spikes, it should trigger an alert through the same on-call pipeline that handles a latency spike. Teams that instrument token counts, tenant IDs, and feature labels from the first production request accumulate attribution data that compounds in value as the portfolio grows.

Token Economics Attribution Flow

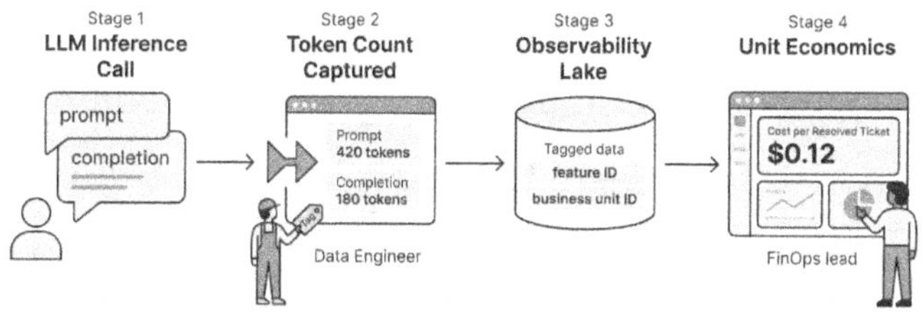

Token tracking enables per-feature unit economics, the foundation for **prompt compression** and model downgrade decisions.

Cost-per-request versus request volume showing token non-linearity decoupling spend from traffic

13.3 Unit-Cost Telemetry: Wiring Cost as a First-Class SLI

The foundation of AI FinOps engineering is measurement. You cannot rate-limit what you have not measured, route to a cheaper model based on a threshold you have not defined, or detect a runaway agent without a cost-anomaly alert. Every optimization and guardrail in this chapter depends on a telemetry layer that surfaces cost-relevant signals at the granularity of individual inference requests and aggregates them into per-tenant, per-feature, and per-task dimensions.

The Core Unit-Cost Metric Set

Cost-per-prediction is the atomic unit. For an API-backed model, it is tokens-in multiplied by the input token rate plus tokens-out multiplied by the output token rate, summed per call. For a self-hosted model, it is GPU-seconds per inference call priced against the instance hourly rate, divided by active batch size. The OpenTelemetry GenAI semantic conventions define standard attribute names: gen_ai.usage.input_tokens, gen_ai.usage.output_tokens, gen_ai.request.model, and gen_ai.system. Adopting these conventions ensures token data from a prompt engineering team, an agent loop, and a batch summarization job all land in the same schema. A tool like Langfuse, LangSmith, or a custom OTel collector can aggregate these spans and compute cost-per-request in real time against a configurable pricing table.

Cost-per-task sits one level above cost-per-prediction. A task is the user-facing unit of work: a document summarized, a support ticket resolved, a code review completed. Linking all spans under a shared trace ID, with the task type as the root span attribute, enables computation of total model spend per task type. If resolving a support ticket costs $0.14 and the queue handles 50,000 per month, the monthly model cost for that feature is $7,000, visible to the feature team in real time rather than discovered at month-end.

Cost-per-tenant is the attribution anchor for multi-tenant platforms. Every LLM gateway request should include a tenant identifier, propagated as a span attribute, through to the token-count metric. Without that tag, cost attribution requires manual log correlation that produces estimates, and per-tenant guardrails cannot be enforced automatically.

Input and output token counts must be tracked as separate metrics because they have different cost profiles and different optimization levers. For most frontier LLM APIs, output tokens cost two to four times as much as input tokens. Track both as histogram metrics. A stable average with a drifting 95th percentile is the early warning sign of a context-stuffing problem before it becomes a billing event.

Cache Hit Rate, GPU Utilization, and Batch Fill Rate

The semantic cache hit rate belongs on the same dashboard as cost-per-request and GPU utilization. A semantic cache with a 40 percent hit rate deflects 40 percent of LLM calls at near-zero cost. If that drops to 15 percent because a prompt template changed without updating the cache key scheme, cost-per-feature climbs by roughly 42 percent. Wire an alert that pages on-call when the hit rate drops more than 10 percentage points below the 7-day rolling baseline for any feature with meaningful inference volume. Label the metric by cache type, model ID, and feature identifier.

For self-hosted inference, the DCGM exporter exposes per-GPU utilization, memory bandwidth, and SM occupancy as Prometheus metrics. Wire these into the central metrics backend alongside request rate and queue depth. A slow-burn alert when 7-day average GPU utilization falls below 50 percent for interactive serving or 70 percent for batch inference catches over-provisioned clusters before they waste another quarter of budget. Plotting KV cache memory utilization alongside batch fill rate and cost-per-token-generated on a single dashboard makes the interplay between memory pressure, throughput efficiency, and cost immediately legible.

Cost Attribution: Chargeback vs Showback

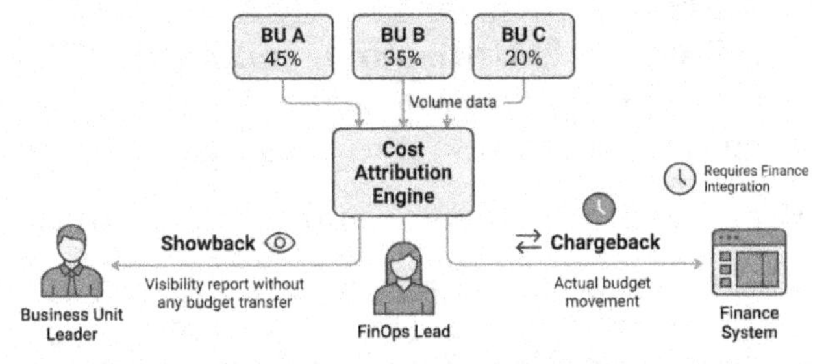

Design for chargeback from initial instrumentation; retrofitting cost tags after deployment is far more expensive.

Unit-cost telemetry dashboard: cost-per-task, tokens-in/out histograms, cache hit rate, GPU utilization, and batch fill rate

13.4 Cost-Saving Engineering Patterns

Measurement reveals where the money goes. The patterns below reduce spend without degrading the product.

Semantic Caching, Response Caching, and Prompt Compaction

A semantic cache stores LLM call results indexed by a vector embedding of the prompt. On each request, the cache performs an approximate nearest-neighbor lookup and returns the cached response if cosine similarity exceeds a configured threshold, without invoking the model. For workloads with high prompt repetition, semantic caching deflects 30-60% of model calls. Start with a threshold around 0.92 and evaluate quality on a sample of deflected requests. Response caching is simpler and appropriate for deterministic prompt patterns. Both cache types require TTL policies tied to staleness tolerance, with hit rate instrumented as a labeled metric.

Prompt compaction reduces tokens-in without changing model behavior. A rolling summary of older conversation turns, rather than the full history, reduces average tokens-in by 40 to 60 percent. System prompt auditing removes redundant clauses; track the average system prompt token count and alert when it increases by more than 20 percent without a deployment event. For RAG pipelines, a reranker filters from 20 chunks to 5 cut context tokens, reducing the number by 75 percent while improving answer quality.

Smaller-Model Routing

A model routing layer evaluates each incoming request and routes it to the smallest model likely to produce an acceptable response. Simple requests (keyword classification, structured extraction from short text, intent classification into a fixed taxonomy) can be handled by a 7B- or 13B-parameter model at 5 to 10 percent of the cost of a frontier 70B+ model. Routing these away from the expensive model while keeping complex multi-step reasoning on the larger model is the highest-leverage single optimization for mixed-complexity workloads.

Implement routing initially with rules (if the request matches a template, use a small model) and graduate to a learned routing classifier once you have labeled data. Wire the routing decision as a metric label on every inference call so the cost dashboard shows, by feature, what fraction of requests are served by each model tier.

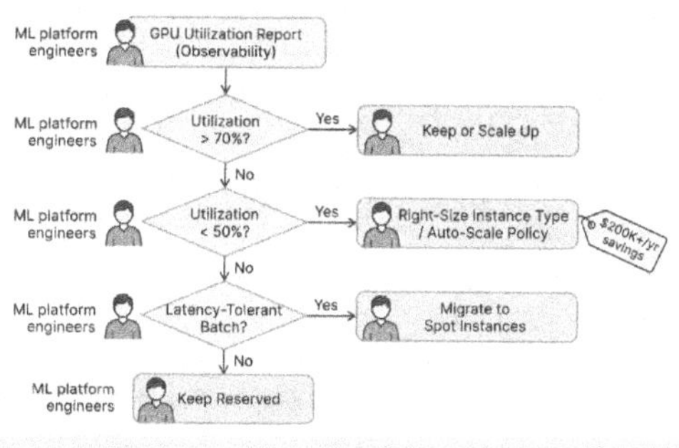

Quarterly utilization review from observability prevents sustained idle GPU billing on over-provisioned clusters.

Model routing architecture: classifier dispatching requests to tiered inference endpoints with per-tier cost tracking

Speculative Decoding, Batching, Autoscaling, and Spot Inference

Speculative decoding uses a small draft model to generate candidate token sequences, which the large target model verifies in a single forward pass. When draft accuracy is high, throughput increases by 1.5x to 3x without changing the output distribution. vLLM, TGI, and TensorRT-LLM support it natively. Track the speculative acceptance rate; below 50 percent signals a poorly matched draft model.

Continuous batching amortizes the fixed costs of a forward pass across concurrent requests, reducing cost-per-request by 85 to 90 percent for high-volume embedding and classification workloads. An autoscaler using predictive scale-out, a minimum warm replica count, and GPU utilization plus queue depth as scaling signals can cut daily serving cost by 30 to 50 percent for endpoints with diurnal traffic patterns. Spot or preemptible GPU instances reduce hourly compute cost by 60 to 80 percent and are appropriate for batch workloads, retraining, and offline evaluation pipelines.

Model Distillation, Quantization, and On-Device Fallback

Model distillation trains a smaller student model to replicate the outputs of a larger teacher, producing a model 4x to 10x cheaper to serve with 5 to 15 percent quality degradation on typical tasks. Quantization reduces the precision of weights from FP16 to INT8 or INT4, shrinking the memory footprint and increasing throughput on the same hardware. Both require evaluation against a task-specific golden dataset before deployment to production. Track the quality delta on the eval harness alongside the cost delta and wire both into the model card. A distilled model that performs within 2 percent of the original on a classification task is an unambiguous cost win.

For applications with mobile or edge clients, running a small on-device model with 4-bit quantization eliminates the need for a cloud API call, enabling latency-tolerant or privacy-sensitive inference. Sub-3B parameter models handle classification, short-text summarization, and intent detection on recent mobile SoCs. Log on-device fallback decisions with the same metadata as cloud routing decisions so the cost dashboard shows on-device versus in-cloud request fractions.

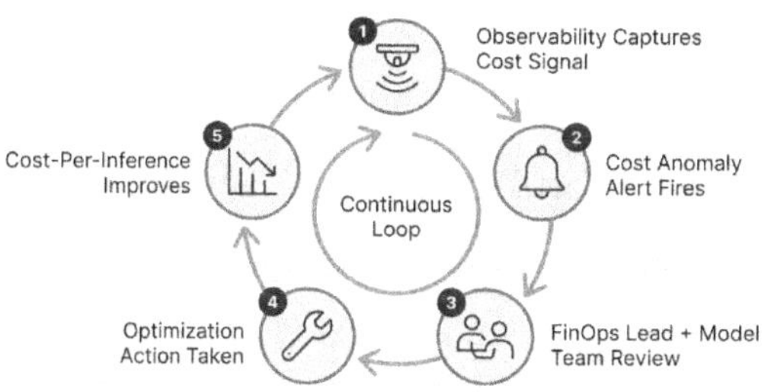

Real-time cost alerts convert periodic FinOps reviews into continuous cost governance for AI infrastructure.

Layered cost-optimization stack: caching, routing, batching, and quantization applied in sequence with per-layer cost-reduction metrics

13.5 Budget Guardrails: Engineering the Cost Control Plane

Cost-saving patterns reduce spend in a steady state. Budget guardrails prevent spend from escaping bounds during incidents, runaway agent loops, and unexpected traffic events. Guardrails are automated mechanisms in the model gateway and agent orchestration layer, not spreadsheet alerts routed to a finance inbox.

Per-Tenant Rate Limits and Token Budgets

Every multi-tenant AI platform should enforce per-tenant rate limits at the model gateway using a token bucket algorithm, with parameters stored in a configuration store that operators can update without a code deployment. When a tenant exceeds their rate limit, the gateway returns a 429 with a Retry-After header rather than silently dropping the request.

Per-tenant token budgets track cumulative consumption against a configured monthly or daily budget. At 80% of the budget, the platform emits a warning event to the tenant's notification endpoint and to the internal on-call dashboard. At 100 percent, the gateway either hard-stops requests or switches to a degraded response mode. The choice must be encoded in configuration, not left to vary by accident. Maintain the counter in Redis, atomically incrementing it by the actual token count returned in the API response. Emit the counter and budget ceiling as metrics so tenant budget utilization is visible in the cost dashboard at all times.

AI Cost Forecast Model Components

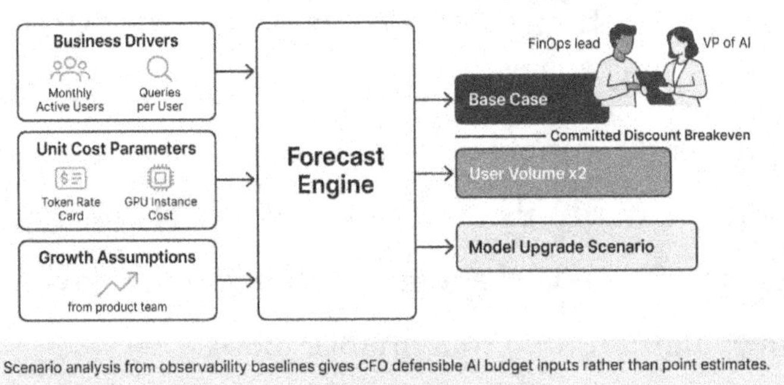

Scenario analysis from observability baselines gives CFO defensible AI budget inputs rather than point estimates.

Per-tenant token budget enforcement: gateway, Redis counter, and metric pipeline with alert thresholds at 80% and 100% of budget

Runaway Agent Kill Switches

Autonomous agents running in a loop can generate superlinear token spend. An agent browsing and summarizing a corpus can call a retrieval tool 400 times in a single session if the termination condition is broken. An agent debugging code can loop indefinitely until a human intervenes or the context window fills. These are documented production incidents at organizations running agentic AI at scale.

The engineering response is a cost sentinel embedded in the agent orchestration layer. The sentinel tracks cumulative token spend and tool call count per session and terminates when either threshold is crossed. If a normal session costs $0.05 to $0.20, a kill threshold of $2.00 gives a 10x safety margin while catching runaway sessions. Log the kill event as a structured incident with full session context and wire it as a high-priority on-call alert. A runaway agent kill is a production incident: the termination condition is broken, the tool is returning malformed responses, or the task prompt is ambiguous enough that the agent cannot recognize success. A soft warning at 50 percent of the session budget allows some agent frameworks to trigger a context-compaction step before the hard limit fires.

Cost-Anomaly Alerts, Routing Policies, and Showback

Cost-anomaly detection uses the same infrastructure as latency-anomaly detection: a baseline window, a threshold, and an alert pipeline. The most useful alerts are: sudden increase in average tokens-in-per-request, sudden decrease in semantic cache hit rate, sudden increase in cost-per-request for a specific tenant, and sudden increase in agent session cost percentile. Route all cost-anomaly alerts to the same on-call rotation that handles latency and availability alerts.

Model-routing policies specify which model tiers are appropriate for each request type and are enforced at the gateway level. Store policies in a version-controlled configuration system with an audit log to track and roll back changes to routing behavior. A routing policy change that inadvertently sends complex tasks to an undersized model surfaces as a quality regression in the evaluation pipeline; the connection between routing configuration and the quality eval dashboard closes that feedback loop.

The showback dashboard should be self-service: any product engineer should be able to look up the cost of their feature for the current billing period without filing a request to the FinOps team. This requires comprehensive feature_id tagging across all LLM gateway spans, accurate cost computation based on actual model pricing, and near-real-time updates. Chargeback wires per-feature cost attribution into the internal billing system, so costs flow to the consuming team's budget. Design instrumentation to support auditability from day one, because retrofitting chargeback-compatible tagging after the fact is significantly more expensive than building it into the initial telemetry schema.

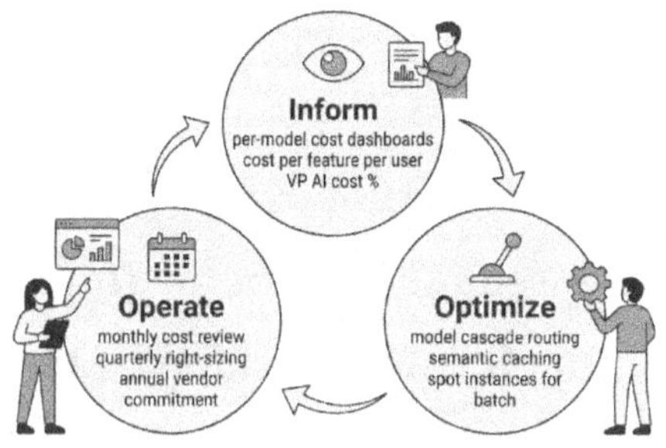

AI FinOps adapts the Inform-Optimize-Operate cycle with token economics and model-version-specific practices.

Budget guardrail architecture: rate limiter, token budget counter, cost-anomaly detector, and routing policy engine layered in the model gateway

13.6 Technical Team Checklist: AI FinOps Engineering

- **Action:** Instrument every LLM inference call to emit gen_ai.usage.input_tokens and gen_ai.usage.output_tokens as span attributes labeled with model_id, feature_id, and tenant_id using OpenTelemetry GenAI semantic conventions. **Owner:** ML Platform Eng. **Done when:** Per-request token counts and cost-per-request are queryable in the observability backend, segmented by feature and tenant, with less than 5-minute latency.

- **Action:** Compute and track cost-per-task for each agent and RAG pipeline by linking all spans in a task session under a shared trace ID with a task_type root attribute and aggregating total token cost per trace. **Owner:** ML Engineer. **Done when:** Cost-per-task-type histogram is available in the cost dashboard and used as the primary cost SLI in each feature's runbook.
- **Action:** Deploy a semantic cache in front of each high-volume LLM endpoint with cache hit rate instrumented as a labeled metric and a page-worthy alert that fires when hit rate drops more than 10 percentage points below the 7-day rolling baseline. **Owner:** ML Platform Eng. **Done when:** the Semantic cache hit rate appears on the same dashboard as cost-per-request, and the alert has been validated in a test scenario.
- **Action:** Implement per-tenant token budget enforcement in the model gateway using a Redis-backed counter with alerts at 80% and 100% of budget and a configurable degraded-mode response at budget exhaustion. **Owner:** ML Platform Eng / SRE. **Done when:** Token budget counters are visible per tenant in the cost dashboard, and a test tenant can hit the budget limit and receive the degraded response without impacting other tenants.

- **Action:** Embed a cost sentinel in every agent orchestration loop that terminates sessions exceeding a configured cost threshold, logs the kill event as a structured incident, and pages on-call with the session trace. **Owner:** ML Engineer. **Done when:** At least one runaway agent simulation has been run in staging, the sentinel has fired correctly, and the alert has been routed to on-call within 60 seconds.
- **Action:** Implement model-routing policies in the gateway configuration system so requests below a token-count or complexity threshold route to a smaller model tier, with routing decisions logged as span attributes. **Owner:** ML Platform Eng. **Done when:** Per-tier inference volume and cost savings from routing are visible in the cost dashboard, and the routing policy is version-controlled with an audit log.
- **Action:** Build a self-service showback dashboard surfacing cost-per-feature-per-day for all LLM-powered features, updated with less than 15-minute latency, accessible to product engineers without a FinOps team request. **Owner:** FinOps Eng / ML Platform Eng. **Done when:** Every feature team with a production LLM integration has reviewed their feature's cost on the dashboard and confirmed the attribution is accurate.
- **Action:** Conduct a quarterly right-sizing review using 90-day GPU utilization and batch fill rate metrics for all self-

hosted serving infrastructure, producing a signed-off action plan. **Owner:** ML Platform Eng / SRE. **Done when:** Any serving cluster with a 90-day average GPU utilization below 50% has a documented right-sizing action assigned to an owner with a deadline.

- **Action:** Wire cost-anomaly alerts covering tokens-in-per-request, cache hit rate, per-tenant spend, and agent session cost into the main on-call alert pipeline alongside latency and error-rate alerts, with runbook links for each. **Owner:** SRE / On-Call. **Done when:** Cost-anomaly alerts have been triggered in a test scenario, responded to by on-call, and the runbook validated against the actual alert context.

Vendor Negotiation Evidence Package

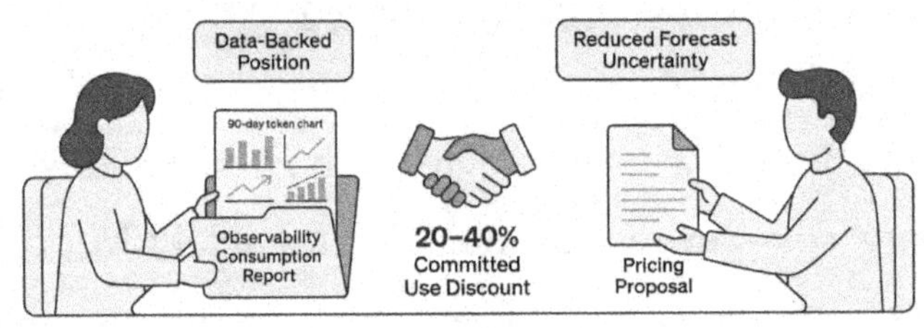

Vendors offer larger discounts to organizations demonstrating consistent, measured consumption trajectories from observability data.

Self-service showback dashboard with cost-per-feature, per-tenant budget utilization, and cost-anomaly alert status

13.7 Team Conversation Starters

- If your LLM endpoint's token cost tripled tomorrow, how long would it take your current alerting setup to page someone, and what would the engineer do first?
- For each LLM-powered feature in production, can you answer: what is the cost-per-task today, and at what user volume does that cost become unsustainable? If not, what instrumentation is missing?
- What is the semantic cache hit rate for your highest-volume LLM endpoint right now? If you do not know, what does that tell you about your cost observability posture?
- If an autonomous agent session enters a runaway loop at 3 AM, what automated mechanism terminates it, and what alert tells the on-call engineer what happened and why?
- Which production LLM workloads have never been evaluated against a smaller or quantized model alternative? What would a 30-minute routing experiment need to show before you would route 50% of those requests to the smaller model?
- When the infrastructure team proposes a right-sizing change to a GPU serving cluster, is that decision made with 90-day utilization histograms and batch fill rate metrics, or from memory of when the cluster was provisioned?

13.8 Key Takeaway

AI cost is a runtime engineering signal that belongs in the same telemetry stack as latency, error rate, and availability. The patterns that make costs manageable (unit-cost telemetry, semantic caching, model routing, per-tenant token budgets, runaway-agent kill switches, and cost-anomaly alerting) are instrumentation and architectural decisions that must be designed into the platform from the first production deployment, not retrofitted after the cloud bill becomes alarming. A platform team that treats cost-per-prediction, cache hit rate, and GPU utilization as first-class SLIs wired into the same alerting pipeline as latency SLOs can detect cost anomalies in hours rather than weeks, route requests to the cheapest capable model tier automatically, and enforce hard limits that prevent a runaway agent or misconfigured integration from generating unbounded spend.

Observability ROI: Cost vs Returns

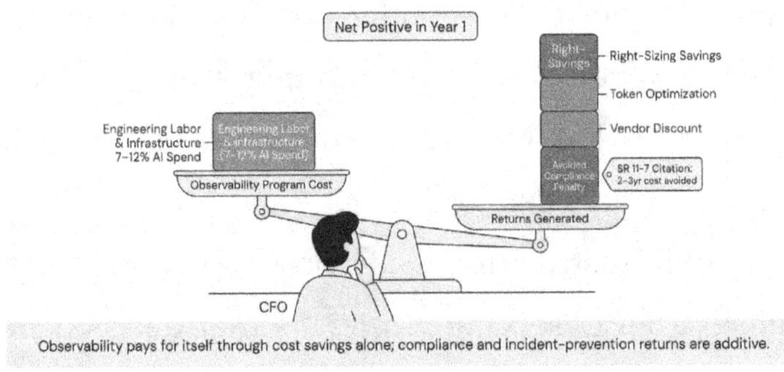

Observability pays for itself through cost savings alone; compliance and incident-prevention returns are additive.

End-to-end AI FinOps engineering stack: telemetry collection, cost dashboard, guardrail enforcement, and anomaly alerting as integrated platform components

14 Building the AI Observability Team

14.1 Opening Scenario

It is a Tuesday morning, and a staff engineer named Priya is pulled into a 30-minute alignment call with her VP of AI/ML. The ask lands flat and heavy: seven production models, no single owner, a state regulator requesting a performance audit in eight weeks, and nobody who can produce a coherent history of model health. The VP wants Priya to define the team that should have existed two years ago and have a working charter by the end of the sprint.

Priya opens a blank doc and starts with ownership boundaries rather than headcount. She sketches five seats: an ML platform engineer who owns the inference infrastructure and the telemetry pipeline; an ML reliability engineer who carries the pager for every model-related alert and owns the runbook library; an eval engineer who owns the golden dataset library and the evaluator-as-judge harness; a governance engineer who owns the audit log architecture and the attestation pipeline; and a FinOps engineer who owns token-cost telemetry and the budget guardrail tooling. Five seats. Five ownership boundaries. No overlap by design.

The on-call model takes fifteen minutes. The ML reliability engineer carries the primary pager for inference latency, error rate, and drift threshold breaches. The ML platform engineer is secondary for infra-layer faults: GPU saturation, collector failures, and model registry unavailability. The eval engineer is on a weekly cadence, not a pager. The governance engineer is async except during an active regulatory inquiry.

The platform service contract fits on one page. Any model team that passes a production readiness review gets SLO coverage, on-call response, drift review on a two-week cadence, and a quarterly eval report. Any model team that skips the production readiness review gets none of those things and cannot claim platform-team incident response. By the end of the sprint, the charter lives in a repository: role definitions, an on-call rotation policy, a production readiness checklist, and the service contract. Priya has converted an organizational gap into a structured engineering function in five working days.

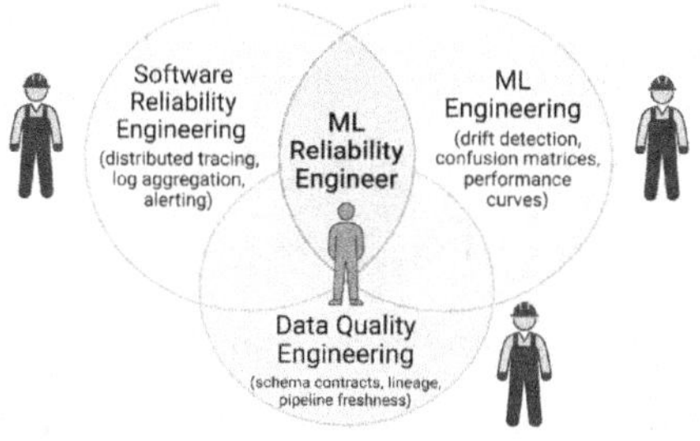

The AI observability skill profile sits at the intersection of three disciplines.

Five-seat AI observability platform team: ownership boundaries and on-call topology

14.2 Why It Matters for Production Systems

The wrong team shape does not produce a single catastrophic failure. It produces a slow accumulation of invisible problems that become expensive all at once. Each failure mode has a clear path from absent ownership to production consequence.

Diffuse instrumentation ownership is the first. When the ML team believes instrumentation is a platform concern and the platform team believes it is an ML concern, no collector gets configured, no telemetry flows, and the first signal of model degradation is a customer complaint. A regional carrier discovered this at eleven weeks and roughly four hundred thousand dollars in audit reconstruction costs, the predictable arithmetic of retroactive observability under regulatory pressure.

The second is no on-call boundary for model failures. A fraud detection model that drifts from 94 percent to 81 percent precision continues to serve predictions with normal latency, achieving a zero percent HTTP error rate. Every application health dashboard reads green. The model silently misclassifies transactions for weeks until a human reviewer notices an anomaly in case outcomes. Without an ML reliability engineer carrying the pager for model quality metrics, that failure mode has no owner and no detection path.

Nobody owns drift. Drift detection requires a reference window, a statistical test scheduled to run against the current inference distribution, and a threshold breach wired into the alerting pipeline. If no engineer is explicitly responsible for scheduling drift reviews, the drift detector either does not exist or exists in a state of quiet staleness, producing false positives or alert fatigue that makes the whole system irrelevant.

No governance engineer is the fourth. The EU AI Act, SR 11-7 in banking, and emerging US state AI regulations require organizations to maintain performance records for high-risk models, with defined retention periods, access controls, and audit-trail immutability. An audit log never designed to be immutable, a model card written once and never updated, a performance history in a data scientist's notebook: these findings produce consent orders and mandatory remediation programs.

The token cost without a FinOps engineer is the fifth. An LLM application making 20,000 requests per day, each with 3,000 tokens, generates 60 million tokens of daily inference spend. Without unit-cost telemetry at the serving layer, the cost signal arrives as a monthly cloud bill line item after the overspend is already sunk. The FinOps engineer instruments token accounting at the request level, builds the budget guardrail that trips a kill switch or throttle at a defined threshold, and produces the unit-cost dashboard showing cost-per-prediction alongside latency and error rate.

The missing hand-off contract is the sixth. A hand-off consisting of a model artifact in the registry and a Slack message is not a hand-off. A production readiness review that gates deployment on documented SLO targets, a reference window definition, a runbook stub, a data schema contract, and a contact list for the originating model team is a hand-off. Without it, every incident starts with a conversation about who is responsible before anyone can talk about what to fix.

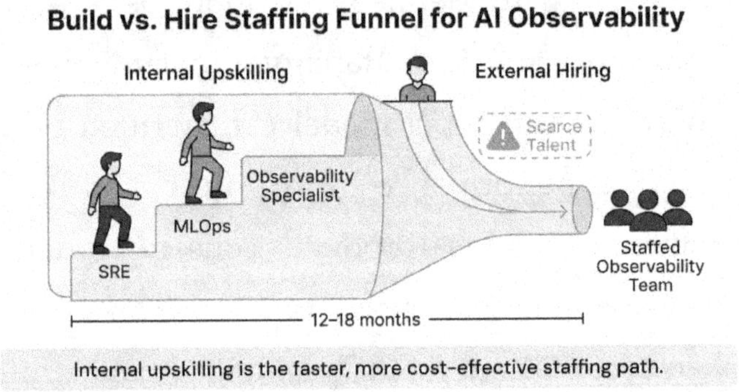

Build vs. Hire Staffing Funnel for AI Observability

Internal upskilling is the faster, more cost-effective staffing path.

Failure mode topology: six paths from missing ownership to production degradation

14.3 Roles and Split of Responsibility

ML Platform Engineer

The ML platform engineer owns the inference serving layer, the model registry, feature store integration, and the telemetry pipeline from the inference endpoint to the observability backend. This role configures OpenTelemetry collectors on the serving layer, ensures trace data carries the span attributes defined by the OpenTelemetry GenAI semantic conventions (model ID, input token count, output token count, request latency, sampling fraction), and maintains the collector infrastructure so telemetry flows reliably to the metrics and trace backends. The skills bar is senior distributed systems plus MLOps depth. A strong hiring signal is whether the candidate can design a telemetry schema for a multi-step LLM pipeline that preserves parent-child span relationships across tool calls and retrieval steps. This role does not own statistical thresholds for drift alerts, the content of the eval harness, or the audit log retention policy.

ML Reliability Engineer and AI SRE

The ML reliability engineer carries the primary pager for model quality in production: drift threshold breaches, eval score regressions, data schema violations at the inference layer, and business-metric anomalies correlating with model behavior. The AI SRE writes the runbooks declaring what to do when a PSI breach fires, how to validate whether a confidence calibration shift is real or an artifact of a stale reference window, when to escalate from an alert to a model rollback, and how to distinguish a model incident from a data incident from an infrastructure incident. An infrastructure incident goes to the ML platform engineer. A model quality incident is routed to the ML reliability engineer, who works with the originating model team's contact to determine whether the drift is pathological or expected. The on-call rotation policy encodes this distinction so the on-call engineer does not have to reason about it under incident pressure. The ML-specific knowledge required covers PSI and KL divergence, precision-recall interpretation, covariate versus concept drift, and how evaluator-as-judge frameworks like RAGAS or DeepEval score faithfulness and groundedness in RAG pipelines.

AI Observability Career Ladder

A defined career ladder retains AI observability talent and builds institutional expertise.

ML reliability engineer on-call decision tree: model incident vs data incident vs infrastructure incident

Eval Engineer

The eval engineer owns the golden dataset library (curated, versioned input-output pairs used at deployment gates and in eval reviews), the evaluator-as-judge harness (the pipeline that scores LLM outputs on faithfulness, relevance, toxicity, and task-specific quality dimensions), and the eval review cadence (a scheduled cycle where eval scores are examined against SLO baselines, regressions are flagged, and decisions are made about retraining or golden dataset updates). The skills bar includes strong Python skills, fluency with eval frameworks (RAGAS, DeepEval, Langfuse evaluations, Evidently for traditional ML), and experience with statistical sampling strategies. The eval engineer is not on a pager rotation, but they are the first person called when an eval score regression triggers a production review.

Governance Engineer

The governance engineer designs and operates the artifact pipeline that turns observability signals into evidence. An attestation record captures, at model deployment time, the model's identity, deployment performance (eval scores, error rate, latency), risk classification, and sign-offs from the model and platform teams. The record is immutable by design: written once to an append-only audit log and retrievable by audit ID at any future point. This role owns the audit log architecture: the storage backend, the event schema, the write pipeline that captures drift events, eval regressions, incident declarations, and model rollbacks, and the retention policy matching regulatory requirements for each model's risk tier. For a high-risk model under the EU AI Act Article 9, that means 10 years of performance history with access controls that allow a regulator to read the log without modifying it.

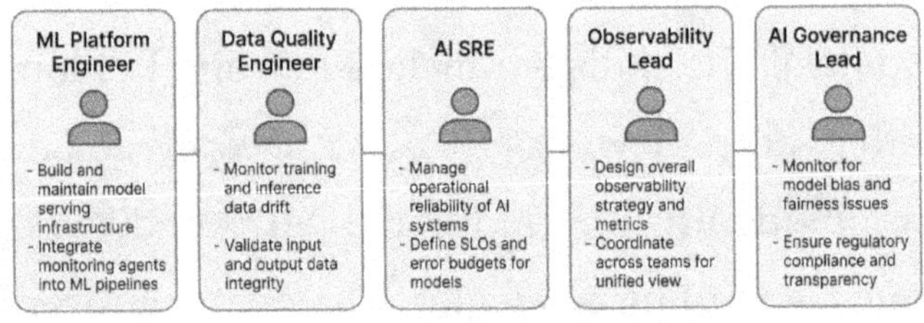

Clear role definitions prevent gaps in accountability across the AI observability program.

Governance engineer artifact pipeline: attestation record lifecycle from deployment to audit query

FinOps Engineer and Platform Architect

The FinOps engineer owns unit-cost telemetry: instrumenting every inference request to record input token count, output token count, model identifier, and calling application identifier as span attributes in the cost analytics backend. They build the unit-cost dashboard showing cost-per-prediction alongside latency and quality metrics, and the budget guardrail tooling that compares rolling spend against a defined budget and either alerts, throttles, or trips a kill switch at defined thresholds. In a founding team, the architect's function is carried by the most senior engineer present. As the team grows, a dedicated architect handles cross-cutting concerns, including integration with the enterprise monitoring stack, schema registry design, distributed tracing topology, and vendor evaluation of tools such as Arize, Langfuse, Datadog APM, and Dynatrace. The architect writes architecture decision records and conducts design reviews before new model teams are onboarded to the platform.

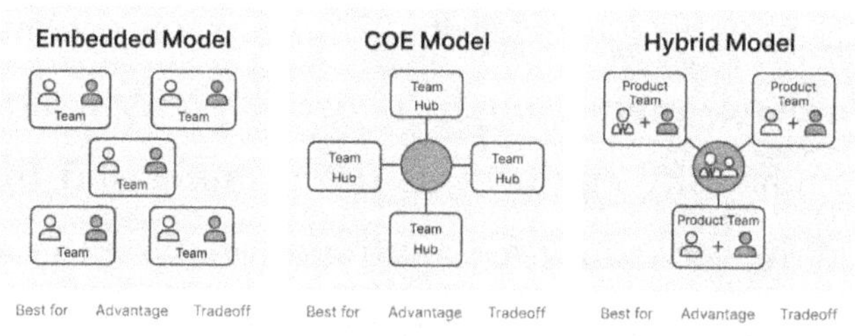

Platform architect scope: cross-cutting concerns, vendor integration topology, and schema registry design

14.4 On-Call and Ownership Boundaries

An on-call rotation is an expression of ownership. For an alert to be actionable, the signal chain must be engineered before the pager fires: a meaningful threshold, a runbook that a non-author can follow, tooling available without escalation, and an escalation path to the originating model team. An alert that fires without those properties is an on-call task.

Who Carries the Pager for What

The on-call ownership matrix breaks down by failure type. Inference infrastructure failures (serving-layer 5xx rate above threshold, collector pipeline failure, GPU node unhealthy, model registry unavailable) go to the ML platform engineer as the primary and the architect as the secondary. Model quality alerts (PSI breach on a monitored feature, KS-statistic over threshold, eval score below SLO target, confidence calibration drift beyond policy bounds) go to the ML reliability engineer as the primary and the eval engineer as the secondary for anything touching the golden dataset or the evaluator pipeline. Cost alerts (rolling spend over budget guardrail, cost-per-prediction above policy threshold, token budget exhausted) go to the FinOps engineer as primary.

The error-budget policy ties the on-call rotation to consequences. Each production model has an SLO for each monitored dimension: inference latency at the 99th percentile, prediction error rate, drift threshold breach frequency expressed as budget consumed per month, and eval score floor. When an error budget is exhausted, the policy defines what must happen: a mandatory architectural review, a freeze on new feature rollouts, or a required retraining cycle. The ML reliability engineer owns the error-budget ledger and produces a weekly report showing budget consumption by model and dimension.

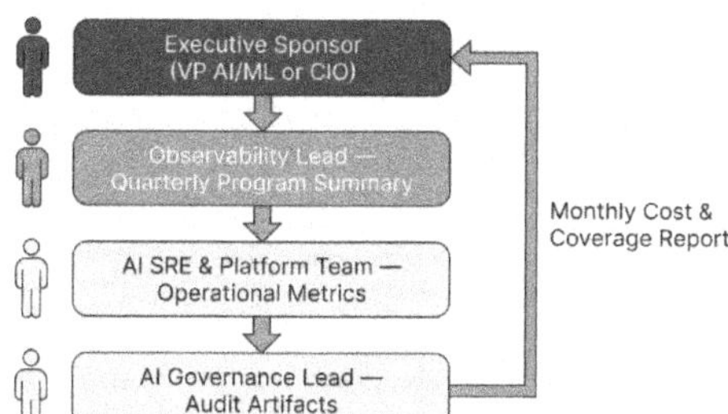

Active executive sponsorship is the strongest predictor of observability program durability.

On-call ownership matrix: failure type to primary/secondary owner with escalation path

The Hand-Off Contract and Ownership Boundary at the Platform-Product Interface

The production readiness review converts an ML team's model from a development artifact to a platform-managed service. The review must verify: the feature schema registered in the schema registry; the agreed SLO targets for latency, error rate, and drift; the reference window definition for drift detection; the golden dataset identifier and version for eval reviews; the runbook stub for the three most likely failure modes; and the escalation contact list naming a specific on-call engineer from the originating team. The platform team's service contract specifies what the model team receives: SLO monitoring with defined alert thresholds wired into the on-call rotation, drift detection on the agreed cadence (two weeks for high-risk models, four weeks for standard-risk), a quarterly eval review with a written output report, incident response coverage during on-call hours, and a monthly cost report for LLM-based models. Models that pass are on the service contract; the on-call rotation does not cover models that do not pass and cannot claim platform-team incident response.

This is where RACI in code becomes literal: the checklist is version-controlled, each item is signed off in a pull request, and the governance engineer uses it as the primary input to the attestation record written to the audit log at deployment. The platform team does not make unilateral rollback decisions for model quality issues. They make unilateral decisions on infrastructure issues: if the serving layer returns 5xx errors, the platform team acts. If the model is producing statistically anomalous outputs but technically serving, the platform team diagnoses and escalates. This boundary must be documented in the on-call runbooks so engineers do not make judgment calls about model correctness under incident pressure.

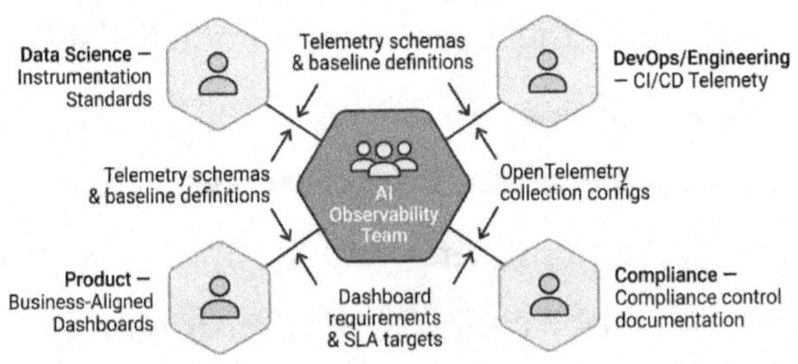

Structured agreements across functions prevent instrumentation gaps and audit surprises.

Platform-product ownership boundary: decision rights at each failure type with escalation paths

14.5 Working Agreements, Design Reviews, and Technical Rituals

Each ritual in this section has a defined cadence, owner, and output artifact. Without them, the platform team's operational clarity degrades to informal communication and undocumented judgment calls.

Production Readiness Reviews and Eval Reviews

The production readiness review is a synchronous event, typically lasting 60 to 90 minutes, chaired by the platform architect or the most senior ML reliability engineer. The originating model team walks through the checklist item by item, the platform team asks clarifying questions, and gaps are recorded as blocking items. Common blockers: the reference window has not been validated; the escalation contact list contains only a team alias; SLO targets are still at placeholder values; or the golden dataset version has not been committed to the eval engineer's repository. The output is a signed-off checklist and a deployment ticket that triggers the governance engineer to write the initial attestation record to the audit log.

The eval review is a scheduled event on a two-week or monthly cadence, owned by the eval engineer. The evaluator-as-judge harness runs against the current golden dataset, scores are compared to the baseline recorded at deployment, and regressions are flagged. The ML reliability engineer correlates eval regressions with runtime drift signals. The output is a versioned report committed to the platform team's repository, which the governance engineer uses as an input to the ongoing audit record. When an eval regression crosses the SLO floor for a high-risk model, the review triggers a mandatory architectural review; the governance engineer records the resulting decision in the audit log.

Drift Reviews, Incident Reviews, and Cross-Team Integration

The drift review runs on the same cadence as the eval review and is owned by the ML reliability engineer. It examines PSI and KS-statistic outputs for each monitored feature, compares the current inference distribution against the registered reference window, and updates the reference windows when the distribution has genuinely evolved. The incident review follows every model incident within five business days: timeline, root cause, contributing factors, what worked in the response, and action items with owners and target dates.

Integration with the security engineering team runs through two touchpoints: participation in production readiness reviews for high-risk models to sign off on input validation and output filtering, and consultancy on the access control layer for the audit log backend. Integration with the data engineering team runs through a shared schema registry: upstream pipeline changes affecting a production model feature trigger a review with the ML platform engineer before the change reaches production, and ingestion freshness alerts are routed to both teams' on-call rotations. Integration with the product engineering team is the most frequent touchpoint; the platform architect conducts a lightweight design review at the start of any feature that significantly changes a production model's input distribution, producing a brief technical note that the originating model team reviews to determine whether the change requires a reference window update, an eval harness update, or a new golden dataset generation cycle.

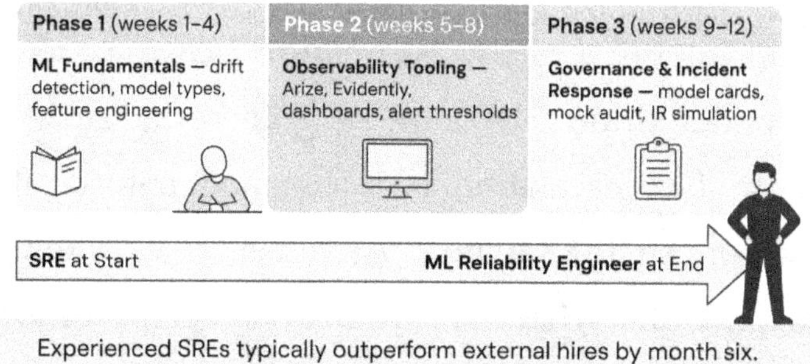

Eval review cadence: input sources, scoring pipeline, output artifacts, and escalation triggers

14.6 Technical Team Checklist: Team Charter and Operating Model

- **Action:** Write the platform team charter defining the five roles with explicit ownership boundaries and a written scope statement for each role, including what each role explicitly does not own. **Owner:** Architect / Staff Engineer. **Done when:** The charter is version-controlled and has been reviewed by the ML platform engineer and the most senior ML reliability engineer.
- **Action:** Define the on-call rotation policy specifying which alert types route to which role as primary and which as secondary, including the escalation path to the originating model team for each alert category. **Owner:** ML Reliability Engineer. **Done when:** The rotation policy is encoded in the incident management tool and has been validated in a tabletop exercise.
- **Action:** Draft the production readiness review checklist with all required artifacts (feature schema, reference window definition, SLO targets, golden dataset version, runbook stub, escalation contacts) and enforce it as a deployment gate. **Owner:** Architect / ML Reliability Engineer. **Done when:** No model has been promoted to production without a completed and signed-off checklist in the prior ninety days.
- **Action:** Design the audit log architecture, including the storage backend, event schema, write pipeline, retention

policy by model risk tier, and access control model. **Owner:** Governance Engineer. **Done when:** The audit log is ingesting events from deployment, drift, and eval review sources, the retention policy matches regulatory requirements for each risk tier, and test queries by model ID and date range return correct results.

- **Action:** Instrument the token cost telemetry pipeline to capture input token count, output token count, model identifier, and calling application identifier on every LLM inference request. **Owner:** FinOps Engineer / ML Platform Engineer. **Done when:** Cost-per-prediction is queryable for all LLM-based production models, and the budget guardrail policy is active with a tested kill-switch or throttle response.
- **Action:** Establish the eval review cadence with a fixed schedule, a defined evaluator-as-judge pipeline, and a versioned output report format. **Owner:** Eval Engineer. **Done when:** The eval review has run on its defined cadence for at least two consecutive cycles, and the output reports are version-controlled and linked from each model's production record.
- **Action:** Run a production readiness review and full onboarding cycle for every existing production model not currently covered by the platform service contract. **Owner:** ML Reliability Engineer / Architect. **Done when:** Every production model has a completed production

readiness checklist, a registered drift monitoring configuration, and a named escalation contact in the incident management tool.

- **Action:** Define the error-budget policy for each monitored SLO dimension and build the budget consumption ledger into the weekly platform team review. **Owner:** ML Reliability Engineer. **Done when:** The error-budget ledger is queryable for all production models, the policy defines mandatory actions at 50% and 100% budget consumption, and the weekly report has run for 4 consecutive weeks.
- **Action:** Conduct a cross-team design review with data engineering to establish the shared schema registry process and the protocol for upstream pipeline changes affecting production model features. **Owner:** ML Platform Engineer. **Done when:** A shared schema registry is in place, and both teams' on-call rotations include ingestion-freshness alerts.

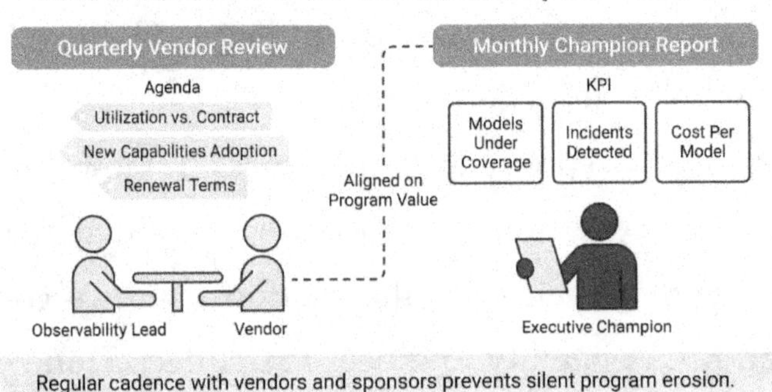

Platform team operating model: ritual cadences, ownership boundaries, and cross-team integration topology

14.7 Team Conversation Starters

- When a production model's PSI breaches the configured threshold at 2:00 AM, who is paged, what is the first diagnostic step in the runbook, and who is the escalation contact at the originating model team if root cause cannot be determined within thirty minutes?
- If the eval engineer discovers that the golden dataset for a high-risk fraud detection model has not been updated in eight months and the production input distribution has shifted significantly, what is the process for updating the golden dataset without creating a gap in the audit record?
- When a product engineering team ships a feature that changes the distribution of a key input feature for a production recommendation model, at what point in the feature development lifecycle should the platform team be involved, and what is the output artifact from that review?
- What is the platform team's service contract with the originating model team for a model that passes the production readiness review, and what happens to that contract if the originating model team does not provide a reachable escalation contact during an active incident?

- If the FinOps engineer's budget guardrail trips a throttle on an LLM application mid-request for a user-visible feature, who owns the incident response, and what is the process for raising the budget threshold before restoring full service?
- How does the governance engineer ensure that the audit log record for a model that was retrained, redeployed, and subsequently rolled back within a single sprint captures all three events with sufficient detail to satisfy a regulatory audit request two years later?

14.8 Key Takeaway

The AI observability platform team is an engineering function, not an oversight committee. Its effectiveness is determined by the precision of its ownership boundaries, the quality of its on-call runbooks, the rigor of its production readiness review process, and the reliability of its artifact pipeline from telemetry to audit log. A five-person team with clearly defined roles, an enforced service contract, and structured rituals for eval review, drift review, and incident review can operate a portfolio of fifteen production models with the kind of operational clarity that makes incidents recoverable and regulatory audits routine. The team shape described in this chapter is not aspirational. It is the minimum engineering infrastructure required to claim that an organization's AI systems are genuinely observable in production, with a named owner for every failure mode and a written record for every decision that matters.

15 The AI Observability Maturity Model

15.1 Opening Scenario

Priya had been the platform architect for the AI infrastructure team for two years when her VP of Engineering pulled her into a conference room and asked a question she could not immediately answer: "Where are we?" A peer organization had resolved a model drift incident in under four hours—a feat that would have taken Priya's team nearly four days. The question was about capability tier: what level of AI observability did the team actually occupy, and what would it take to close the gap?

Rather than guessing, Priya ran a structured two-day capability assessment. She pulled together the on-call SRE lead, two ML platform engineers, and the data quality engineer, then scored the program across thirty dimensions: telemetry coverage, alert fidelity, eval harness completeness, drift detection coverage, governance artifact generation, FinOps instrumentation, runbook quality, and replay capability. For each dimension, she demanded artifact evidence—not "we think we do PSI-based drift detection," but "show me the alert history from last Tuesday when the credit model's feature distribution shifted."

The resulting heatmap was uncomfortable. On infrastructure telemetry—GPU utilization, endpoint latency, and API error rate—the team scored Level 2. Their

OpenTelemetry pipeline was solid, Prometheus was scraping every inference endpoint, and Grafana dashboards covered every serving cluster. On eval coverage, the score was Level 0. No automated evaluator existed. No golden dataset had been assembled. No eval harness was wired into the deployment pipeline. Model quality post-deployment was inferred from the business KPI lag—exactly the blind spot the peer organization had eliminated.

The heatmap revealed a program instrumented for infrastructure health and operationally blind to model quality. That asymmetry was the answer Priya brought back: the team was not at any uniform tier. They were a patchwork, with patches carrying different risk profiles. The maturity model gave them a shared vocabulary and a sequenced roadmap. The conversation shifted from "we need to do better" to "here are six capability dimensions at Level 0 or Level 1, ranked by blast radius, each with an owner and an acceptance criterion."

Level 1 vs Level 2 AI Observability Detection Gap

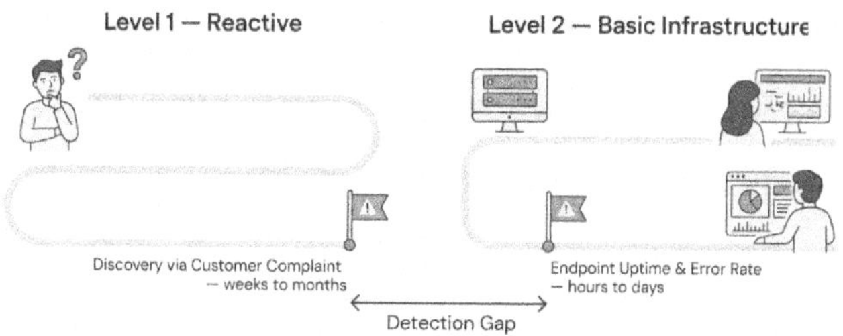

Moving from Level 1 to Level 2 is a technology investment; Level 2 to Level 3 is organizational.

Capability heatmap from a two-day maturity assessment: infrastructure telemetry at Level 2, eval coverage at Level 0, revealing the most actionable engineering gap

15.2 Why It Matters for Production Systems

Ad-hoc AI observability roadmaps share a common failure mode: they are organized around tool purchases rather than capability dimensions. A team acquires a drift detection platform, deploys it against two high-priority models, and declares the problem solved. Six months later, they are paged because a third model — never onboarded to the detector — silently degraded for three weeks. The tool was correct. The capability was never built. A capability model makes this distinction operational: it defines not just what the team can do but what it routinely does, how it knows it is doing it reliably, and how it detects when coverage has eroded. Without that frame, telemetry gets wired only for systems active during a funded project, eval harnesses exist only for models whose ML engineer was particularly rigorous, and governance artifacts appear only for models that recently faced an audit. Everything else is Level 0, whether or not the team has noticed.

The consequences of uneven coverage are asymmetric. A Level 0 capability in infrastructure telemetry means the on-call engineer cannot distinguish between a model-quality problem and a serving-infrastructure failure. A Level 0 capability on eval means quality regressions from prompt template edits, fine-tuning runs, or dependency bumps propagate undetected until a business KPI drops. A Level 0 capability in governance attestation means that every compliance review requires a manual artifact-assembly sprint, with answers that may be wrong. Production AI systems are continuously modified through retraining pipelines, retrieval index refreshes, embedding model upgrades, and dependency version bumps — each of which can be a potential regression vector. Detecting regressions across that change surface requires eval and drift coverage that fires automatically on every change event. A classification model that degrades from 91 percent accuracy to 78 percent will eventually surface via a downstream KPI alert. Still, MTTD from root cause to first alert is often measured in weeks at Level 0 on eval.

The sequencing problem compounds everything. Teams that invest in autonomous remediation before reliable detection are wiring a closed-loop control system to broken sensors. You cannot self-heal what you cannot reliably detect. Teams that build eval harnesses before stable data-quality pipelines find that golden datasets go stale. Teams that automate governance artifact generation before the model registry is the authoritative source of deployment state find that the automation produces documents that do not reflect what is running in production. A capability model with explicit level dependencies prevents this class of sequencing anti-pattern by surfacing prerequisites before investment is committed.

The on-call argument is the most concrete. A Level 2 team on model monitoring can see infrastructure state but not model quality state — resolving a quality incident at Level 2 typically means waking ML engineers not on rotation, because the SRE lacks the signals to determine whether the root cause is in the model or in the upstream data pipeline. Every capability tier improvement directly reduces MTTR for AI-specific incidents. A capability model converts this abstract goal into a queryable signal: instead of "are we doing AI observability?" the team asks "what is our score on eval coverage, and what does it take to move that score from Level 1 to Level 2?" The second question generates a sprint backlog with owners and acceptance criteria.

Level 3 to Level 4 AI Observability Capability Jump

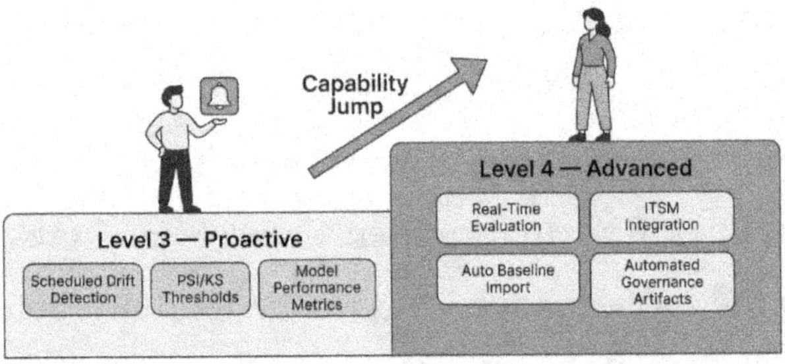

Level 4 is defined by cross-functional integration, not just additional tooling.

MTTD for model quality regressions by capability tier, showing the order-of-magnitude reduction in detection lag between Level 0 and Level 4

15.3 Five Capability Tiers: Engineering Signals at Each Level

The five-tier ladder maps engineering-observable state at each level, defined by what telemetry exists, what alerts fire, what eval runs, what governance is generated automatically, and what cost guardrails are active. A team occupies the tier in which it can produce artifacts that satisfy all criteria for that tier across a given capability dimension.

At Level 0, no intentional AI observability exists. Telemetry is incidental—cloud provider defaults or a Prometheus installation configured for non-AI workloads. No model-specific spans are emitted, no feature distribution baselines exist, no model quality alerts fire, and no eval harness runs. If the team learns about model quality issues from users or from a business analyst noticing a KPI drop, the team is at Level 0 regardless of the tooling available. MTTD for quality regressions is measured in weeks. On-call posture covers infrastructure only; anything short of a hard endpoint failure sits outside the alert space. Level 1 establishes infrastructure telemetry: an OpenTelemetry collector deployed against each model serving endpoint, with latency (p50, p95, p99) and error rate collected at five-minute resolution and retained for thirty days, and alerts on hard infrastructure thresholds. The telemetry stack treats the model as a black box — it measures whether the box responds, not whether its responses are correct or calibrated. No eval harness exists. FinOps at Level 1 is cost-center accounting: total AI compute spend is visible but not broken down by model, endpoint, or business unit.

Level 5 Predictive AI Observability Loop

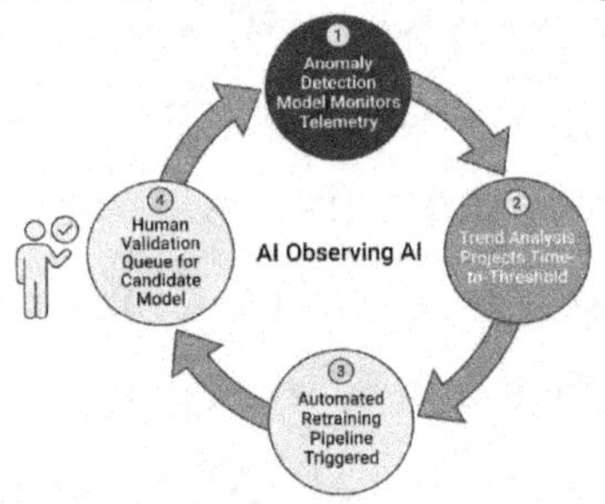

Level 5 delivers forward-looking signals and partially automated remediation.

Telemetry coverage matrix at Levels 0 through 2: what is instrumented, which alert conditions exist, and where model quality signal gaps remain

Level 2 adds model quality signals on top of Level 1 infrastructure telemetry. Feature distribution baselines are computed from training data or a representative reference window. PSI and KS statistics run on a scheduled cadence (daily for high-risk models, weekly for others) and alert when PSI crosses model-specific thresholds (0.2 for warning, 0.25 for critical). SLOs cover both infrastructure and model quality dimensions (drift coverage ratio, eval pass rate). Error budgets are tracked; when one falls below 20%, a health review is required before further deployment changes proceed. Eval coverage at Level 2 is partial: an offline eval harness runs weekly against a golden dataset for the highest-risk models, but is not yet a deployment gate. FinOps at Level 2 adds per-model cost attribution: token spend and compute cost per request are recorded in a queryable cost ledger.

Level 3 makes eval a deployment gate: a new model version cannot be promoted unless it passes the harness against the golden dataset. Eval coverage ratio (platform SLO minimum 0.8) and drift coverage ratio (platform SLO minimum 0.9) appear in the weekly operational review. Every model in the registry carries a monitoring profile with model-specific thresholds, a defined reference window, and alert routing rules. Agent tracing at Level 3 is first-class: agentic workflows emit structured spans for each tool call, LLM invocation, and retrieval operation, with span linking providing end-to-end latency attribution and prompt template versions in span metadata. Governance at Level 3 is automated for the deployment path: model cards are generated by the deployment pipeline, validation reports are produced by the eval harness as build artifacts, and attestation completeness is a governance SLO. FinOps at Level 3 adds token-level accounting with budget guardrails: 80% of the monthly budget (triggering a review) and 95% (triggering a capacity throttle).

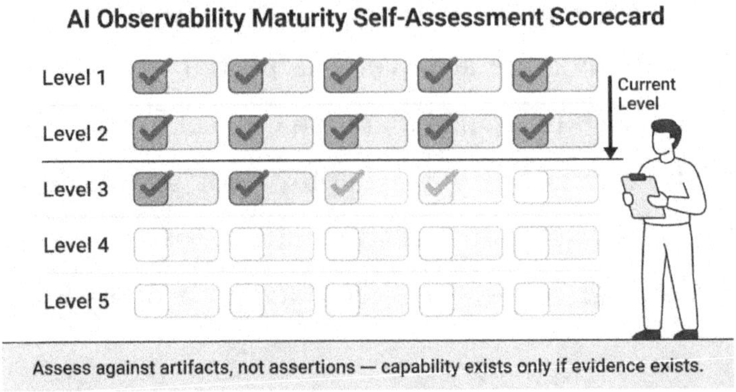

AI Observability Maturity Self-Assessment Scorecard

Assess against artifacts, not assertions — capability exists only if evidence exists.

Eval harness integrated as a Level 3 deployment gate, blocking model version promotion on eval failure, and writing results to the model registry as build artifacts

Level 4 closes the control loop. When a model's PSI score crosses the critical threshold, an automated workflow fires: the drift event is logged to the incident management system, a retraining pipeline is triggered with the current production distribution as the new reference window, and the on-call engineer receives a structured notification with drift statistics, the affected feature set, and the triggered remediation action. Autonomous remediation is scoped to well-characterized failure modes: rolling back a model version when a promotion's eval score drops below a threshold, scaling down an underutilized endpoint when cost-per-request exceeds a budget guardrail, and triggering a cache warm after a retrieval index refresh. The sequencing rule is absolute: do not invest in autonomous remediation before Level 3 detection reliability is established. A documented false-positive rate below 5% over a 90-day window is the prerequisite for wiring any alert type to automated action. MTTD at Level 4 is measured in minutes for instrumented failure modes. FinOps at Level 4 wires a cost sentinel to the token spend ledger: when per-request cost rises more than 30% above a model's 30-day rolling average, the sentinel fires, and endpoint rightsizing recommendations are generated automatically from utilization telemetry.

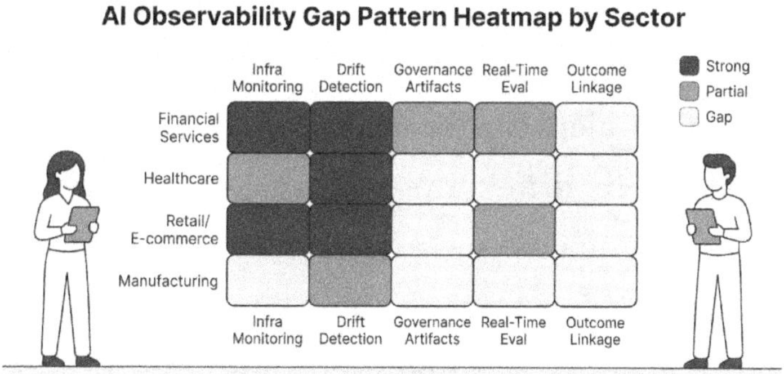

AI Observability Gap Pattern Heatmap by Sector

Sector-specific gap patterns help calibrate roadmaps against peer benchmarks.

Closed-loop control architecture at Level 4: signal path from drift detector through remediation orchestrator to model registry and back to monitoring

15.4 Capability Dimensions: Scoring the X-Axis

The five tiers describe vertical progression. The capability dimensions describe the horizontal scoring surface: thirteen engineering areas, each assessed independently, because a team's tier is rarely uniform across all of them. A production AI program can score Level 3 on infrastructure telemetry and Level 0 on replay capability at the same time. The per-dimension heatmap is the program's actual state, almost always more informative than any aggregate number.

Thirteen capability dimensions form the X-axis of the scoring heatmap. For each dimension, identify the highest tier where all criteria are met by artifact evidence and score the dimension there.

Infra observability progresses from cloud provider defaults (Level 0) through OpenTelemetry-based telemetry with 30-day retention and alerting (Level 1), per-model SLOs with error-budget tracking (Level 2), auto-generated runbooks tied to alert conditions (Level 3), and automated remediation with drift-correlated root-cause attribution (Level 4). Data quality instrumentation covers inference inputs at the same five tiers: no validation (Level 0), schema validation (Level 1), feature-level null and cardinality monitoring (Level 2), a data quality SLO (Level 3), and automatic upstream pipeline investigation (Level 4). Model monitoring covers behavioral signals: no output distribution tracking at Level 0, prediction confidence histograms at Level 1, scheduled PSI and KS statistics at Level 2, a platform-managed service covering all registry models at Level 3, and monitoring signals feeding automated remediation at Level 4.

Eval coverage ratio is the fraction of production models covered by an automated evaluator: none at Level 0; a golden dataset and offline eval script for at least one model at Level 1; scheduled cadence for a subset at Level 2; deployment gate with evaluator-as-judge for LLM systems at Level 3; continuous coverage with replay-based eval triggered on drift at Level 4. Attestation completeness (fraction of models with a current harness-generated eval report) complements eval coverage ratio: a team at Level 3 on coverage but Level 1 on attestation has automation that is not producing durable compliance records -- fix the artifact pipeline before expanding coverage breadth.

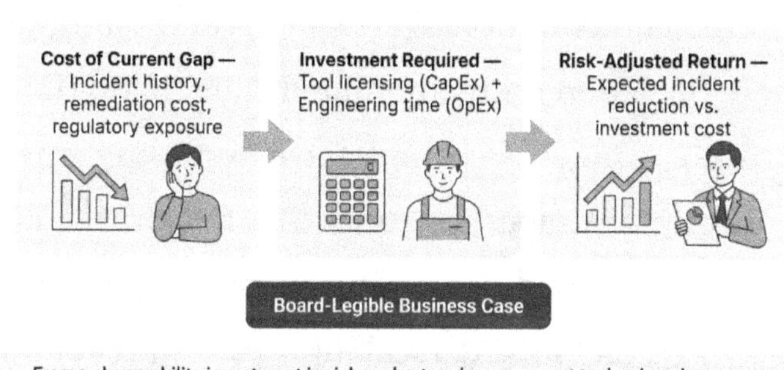

Frame observability investment in risk-and-return language, not technology language.

Eval coverage ratio and attestation completeness as platform SLO metrics, per-model breakdown exposing coverage gaps and stale attestation records

Agent tracing covers multi-step agentic workflows: no structured telemetry at Level 0; basic completion logs at Level 1; per-span attributes for each tool call and LLM invocation at Level 2; full linked traces with prompt template version in span metadata at Level 3; anomaly detection on loop count, session cost, and tool call failure rate with automated circuit-breakers at Level 4. Governance instrumentation progresses from manual assembly on request (Level 0) through templates and documented process (Level 1), deployment-time artifact production relying on engineer discipline (Level 2), pipeline-generated build artifacts with attestation SLO (Level 3), and full automation with immutable audit logs for all model state transitions (Level 4). FinOps instrumentation progresses from cost-center-only visibility (Level 0) through per-model compute attribution (Level 1), per-request cost ledger (Level 2), cost SLOs with budget alerts and auto-generated rightsizing recommendations (Level 3), and autonomous cost anomaly detection with automated rightsizing (Level 4). On-call posture ranges from infrastructure-only coverage (Level 0) through AI-specific runbooks for drift and eval failure with documented escalation criteria (Level 2), ITSM integration with SLO-linked escalation paths (Level 3), and structured decision trees with automated pre-investigation context (Level 4). Runbook quality is scored separately: Level 3 requires dashboard links, query examples, and escalation decision criteria; Level 4 requires quarterly game-day validation. Replay capability: Level 2 has a replay harness; Level 3 automates replay on drift alert to quantify the regression; Level 4 integrates replay as a validation step before promotion. Reference architecture standardization: Level 2 enforces the reference architecture for new deployments via the pipeline gate; Level 3 migrates existing models on a documented schedule; Level 4 evolves the reference from operational learnings and actively refactors legacy deployments.

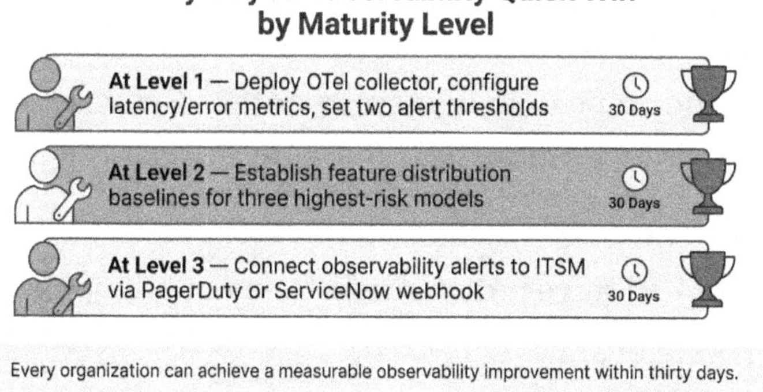

Capability dimension scoring heatmap for a representative production AI program, flagging highest-risk gaps for prioritized roadmap investment

15.5 Practical Roadmap Mechanics: Scoring, Cadence, Anti-Patterns, and Sequencing

Score each capability dimension on the 0-4 integer scale. Identify the highest tier where all criteria are demonstrably met by artifact evidence -- a team member can produce the artifact within five minutes. For drift detection at Level 2, the artifact is an alert history that shows drift alerts fired and investigated in the past 90 days. For Level 3 eval coverage, it is a pipeline log showing the harness gate ran across the last three promotions. For Level 3 governance, it is a pipeline-generated model card for the most recently deployed model version. Where an artifact cannot be produced, the capability scores at the tier below the unmet criterion. Partial scores (for example, 2.5) apply when criteria are partially met. Version the scoring rubric and record the rationale whenever it changes. Run a full baseline assessment at program initiation and repeat every six months. Between formal assessments, track three leading indicators weekly: eval coverage ratio, drift coverage ratio, and attestation completeness. Trigger dimension re-evaluations between cycles when relevant events occur.

Five anti-patterns recur in AI observability programs. Tool-ownership: a team deploys a monitoring platform against three of twelve models and counts the deployment as Level 3 drift coverage. The actual drift coverage ratio is 25 percent, which maps to Level 1. Score against coverage ratios, not tool deployments. Assessment theater: "We do drift detection" is not evidence; a scheduled job that runs PSI comparisons, writes results to a metrics backend, and has alert thresholds with documented escalation paths is evidence. Big-bang roadmap: targeting all dimensions for a two-tier jump simultaneously produces partial progress everywhere and full compliance nowhere -- bring the highest-priority dimension to the next tier completely before expanding scope. Over-automation: investing in Level 4 autonomous remediation before Level 3 detection reliability is established; a drift detector with a false-positive rate above ten percent should never trigger automated action. Coverage stagnation: failing to onboard new models as the portfolio grows. The countermeasure is a deployment gate requiring a monitoring profile, an eval harness registration, and a governance artifact template before any model reaches production.

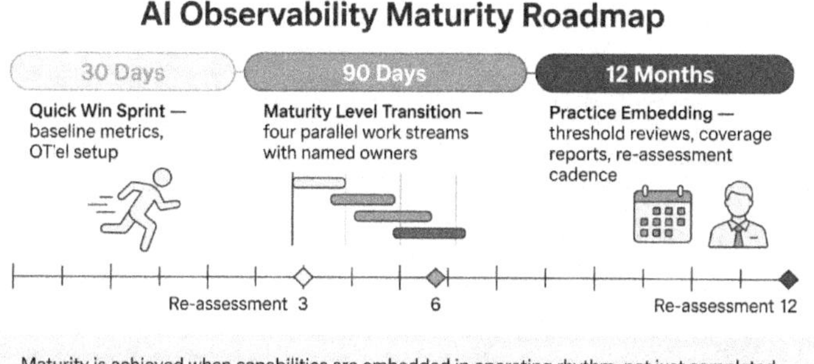

Maturity is achieved when capabilities are embedded in operating rhythm, not just completed.

Anti-pattern dependency map: five common sequencing failures in AI observability maturity programs and the capability preconditions each one violates

Six sequencing rules encode capability dependencies that, when violated, cause investment to fail or underperform. Do not configure eval before infrastructure telemetry is stable and queryable -- eval results are actionable only when correlated with infrastructure state, model version, and data pipeline state. Do not build golden datasets before production input logging is in place — synthetic data does not capture the actual distribution of production inputs. Do not automate governance artifact generation until the model registry is the authoritative source of deployed model state; models promoted outside the registry produce inaccurate governance documents. Do not set SLO-linked budget alerts before per-request cost attribution is in place — aggregate alerts identify overspend but not the responsible model. Do not invest in autonomous remediation until the detection signal has been validated for 90 days, with a documented false-positive rate below 5%. Do not expand coverage breadth before validating coverage depth on a pilot set of three to five models -- the pilot validates tooling, process, and rubric before the full portfolio rollout.

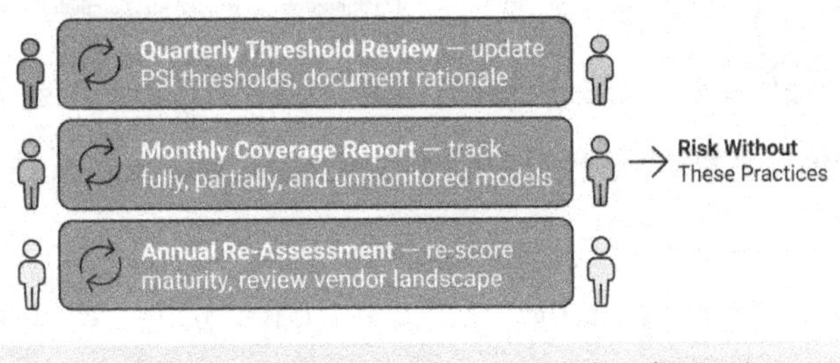

Capability dependency graph encoding the six sequencing rules as directed edges, with anti-pattern violations shown as edges that bypass required preconditions

15.6 Technical Team Checklist: Maturity Assessment and Roadmap

- **Action:** Run a baseline capability assessment against all thirteen dimensions using artifact evidence for every production model. **Owner:** ML Platform Eng. **Done when:** Every dimension has a documented score from 0 to 4, with the specific artifact or artifact gap that determined the score, and the score is reviewed and signed off by a second engineer who did not build the capability.
- **Action:** Track eval coverage ratio and drift coverage ratio as platform SLO metrics with weekly reporting and defined error budget targets. **Owner:** ML Platform Eng. **Done when:** Both metrics are visible in the operational dashboard, with SLO targets and error-budget tracking active.
- **Action:** Instrument every new model deployment with minimum Level 1 observability as a deployment pipeline gate that blocks promotion if a monitoring profile, eval harness registration, and governance artifact template are absent. **Owner:** ML Platform Eng. **Done when:** The CI/CD pipeline fails the build when no monitoring profile exists in the observability platform.
- **Action:** Wire high-priority drift alerts into the ITSM incident workflow with severity classification and SLO-linked escalation paths to the AI on-call rotation. **Owner:**

SRE / On-Call. **Done when:** A synthetic drift event triggers an incident record in the ITSM system that routes to the AI on-call within five minutes of the alert firing.

- **Action:** Write or update runbooks for each known AI failure mode (drift event, eval failure, data quality incident, cost anomaly) with diagnostic steps, dashboard links, and escalation criteria. **Owner:** SRE / On-Call. **Done when:** Each runbook has been exercised in a game-day in the past ninety days, and gaps have been closed.
- **Action:** Integrate the eval harness into the model deployment pipeline as a promotion gate and write eval results as build artifacts to the model registry. **Owner:** ML Engineer. **Done when:** Three consecutive model promotions have been gated by the eval harness, and eval results are queryable from the model registry alongside the model artifact.
- **Action:** Establish per-request cost attribution for all LLM-based systems by recording input token count, output token count, model version, and business unit as structured log fields on every inference request. **Owner:** FinOps Eng. **Done when:** The cost ledger produces a per-model, per-business-unit, per-day cost breakdown for the past ninety days without manual data assembly.
- **Action:** Stand up a quarterly maturity review cadence that re-scores all thirteen dimensions, compares against

the prior baseline, and produces a sequenced roadmap that respects the no-skip dependency between detection, eval coverage, and any autonomous remediation work. **Owner:** Architect. **Done when:** Two consecutive quarterly reviews have been completed with delta heatmaps published to the platform team, and at least one anti-pattern (premature autonomy, unbalanced investment, or detection without response) has been blocked from the roadmap with documented rationale.

15.7 Team Conversation Starters

- When we say a model is "monitored," what specific artifacts can we produce within five minutes to prove the monitoring is operational rather than just configured at some point in the past?
- Which of our thirteen capability dimensions has the largest gap between the score we would self-report without evidence and the score we can back with artifacts? What does that gap reveal about where investment is genuinely needed?
- Our eval coverage ratio is a number. What is it today versus six months ago, and what is the trend? If the portfolio is growing faster than the coverage ratio, what deployment gate policy reverses that trend?
- If our drift detector fires at 2 AM with a PSI critical alert on the credit scoring model, can the on-call engineer

diagnose and initiate remediation without waking up an ML engineer? Walk through the runbook step by step.

- We are planning to invest in automated remediation for drift events. What is our documented false-positive rate for drift alerts over the past ninety days, and does it meet the five-percent threshold required before automation should act on that signal?

15.8 Key Takeaway

The AI observability maturity model is a tool for identifying and measuring coverage gaps, such as eval and drift ratios, attestation completeness, and runbook quality. It guides teams to focus on detection before remediation and ensures discipline in platform investments, reducing costly mistakes. Teams that regularly assess and update the model based on current portfolio and tooling maintain effective AI observability aligned with real production risks.

16 The Road Ahead: Observability in the Age of Autonomous AI

16.1 Opening Scenario

Priya Shankar, principal engineer on the AI platform team at a mid-size fintech, has been handed a blank whiteboard and one mandate: design the next-generation observability stack for the company's autonomous multi-agent settlement pipeline. The existing stack was built eighteen months ago for single-model inference endpoints. That world no longer exists. The pipeline now runs seven collaborating agents: a document-extraction agent, a rule-evaluation agent, a fraud-pattern agent, a counterparty-lookup agent, an exception-routing agent, a reconciliation agent, and an orchestrator that sequences and retries the others. Together, they process 3,400 settlement events per day without human sign-off on any individual decision.

The existing observability stack monitors the infrastructure: Kubernetes pod CPU usage, container restarts, and and p99 latency at the outer API gateway. It sees nothing inside the agent graph. When a settlement event traverses six agents before failing at reconciliation, the trace that lands in the logging backend is a single HTTP 404 with no parent context linking it to the six prior spans. Priya can confirm that something failed. She cannot reconstruct why the error occurred, which agent introduced it, or whether the failure pattern is systematic.

The architecture she sketches over the next hour treats the observability platform as an agent. It subscribes to the structured trace stream from the pipeline and maintains a behavioral fingerprint for each agent type. When a fingerprint deviates beyond a learned threshold, it opens the runbook library, selects a remediation pattern, and authors a draft runbook entry if no match is found. When a prompt template change correlates with a quality regression across three or more sessions, it files a pull request against the prompt template repository with revert evidence attached. When a specific agent's error rate crosses the kill-switch threshold, it emits a workload-pause signal that the orchestration layer enforces, suspending new work assignments while in-flight sessions complete.

None of this requires unavailable technology. What has been missing is the architecture that wires these capabilities together and treats the observability platform as an active participant in system control rather than a passive recorder of state.

Model-Centric vs Outcome-Centric AI Observability

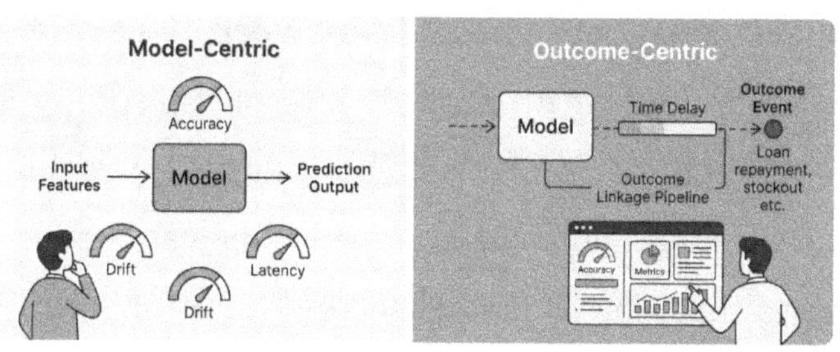

Outcome observability closes the gap between prediction quality and real-world decision impact.

Architecture sketch for observability-as-agent: behavioral fingerprint subscriptions, runbook authoring tool, PR-filing tool, and kill-switch signal bus wired to a seven-agent settlement pipeline

16.2 Why It Matters for Production Systems

The observability primitives built over the past decade were designed for synchronous microservices where causality flows in one direction: a request enters a bounded call graph and returns a response. This model breaks down in ways that are architecturally disqualifying for autonomous, multi-agent, continuously-learning systems. Five distinct failure modes make the gap structural rather than configurable.

Causal opacity across agent boundaries is the first failure. A traditional distributed trace assumes span context can propagate via standard HTTP headers between services under the same instrumentation discipline. In a multi-agent architecture, an orchestrator may delegate to a third-party agent API that does not honor the traceparent header, or to an in-house sub-agent whose framework does not carry OpenTelemetry context through its internal reasoning loops. The trace covers the HTTP call boundary but is blind to everything the sub-agent did to produce its response.

Temporal mismatch is the second failure. Attributing quality regressions to specific model versions, prompt template revisions, or retrieval index updates requires that every telemetry record include the exact artifact version vector active at inference time. Current practice is inconsistent: teams log the model name but not the model content hash, the template name but not its content hash. Version ambiguity turns root cause attribution into manual investigation.

The missing outcome layer is the third failure. Conventional observability measures what the system did. An autonomous procurement agent that executes twelve tool calls and generates a purchase order has not produced a good outcome if the order violates vendor policy. Without outcome observability, the pipeline appears healthy when it is not. The absence of closed-loop control is the fourth failure: the monitoring-alert-human-response loop is too slow for autonomous systems at volume. The fifth failure is telemetry scale: a pipeline with 100 concurrent sessions, averaging 40 reasoning steps each, generates 4,000 spans per second at the reasoning layer alone, requiring intelligent sampling using behavioral signals to preserve full traces for anomalous sessions while retaining only statistical summaries for the population.

16.3 Cross-Agent Causality and Trace Stitching

Reconstructing why a multi-agent system produced a specific outcome requires a complete causal graph of every decision, delegation, and tool invocation that contributed to it. Assembling that graph when agents span organizational boundaries and communicate through message queues is the central hard problem of agentic observability.

Span Linking Across Organizational and Tenant Boundaries

The W3C Trace Context standard propagates trace and span identifiers via the traceparent header and breaks down into three common multi-agent scenarios: third-party agent APIs that do not honor it; queue-mediated handoffs where message bodies carry no standard context field; and multi-tenant pipelines where trace data must be isolated by tenant while still supporting intra-tenant causal graph assembly.

For third-party agents, the solution is explicit span linking using OpenTelemetry's SpanLink API. When the orchestrator receives a response containing a correlation identifier, it creates a span link from its current span to that identifier. If the vendor provides a trace export mechanism, those spans can be ingested and associated with the linked identifier. For queue-mediated handoffs, serialize W3C traceparent into a message attribute at produce time and extract it at consume time using OpenTelemetry messaging instrumentation. OpenTelemetry's semantic conventions for messaging systems define the attribute names, and both Kafka and SQS instrumentation libraries implement the pattern.

Federated traces are assembled from span data collected across multiple independent trace stores with per-store access control. When span context propagation fails, behavioral fingerprinting provides a probabilistic fallback for correlation. A behavioral fingerprint is a feature vector computed from observable execution properties: tool call sequence, inter-tool latency distribution, output token count distribution, and decision confidence score pattern. An offline correlation engine computes similarity scores between orchestrator and sub-agent session fingerprints across overlapping time windows, achieving 85 to 95 percent attribution accuracy in controlled evaluations.

Multi-Agent Trace Propagation Graph

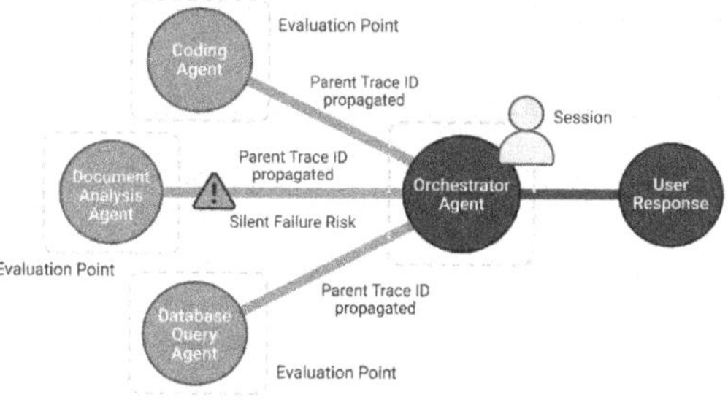

Evaluation at every delegation boundary is required to detect silent multi-agent failures.

Span linking diagram for a multi-agent pipeline: W3C traceparent propagation across HTTP, SpanLink records for third-party agent calls, and message-attribute context propagation across Kafka handoffs

Behavioral Fingerprinting and Replay-as-Test

Behavioral fingerprinting also serves anomaly detection. A fingerprint computed over a rolling window of recent sessions is compared to the agent's historical baseline using cosine similarity or Mahalanobis distance. A sudden shift, such as a sharp increase in a specific tool call type, an unusual distribution of token counts, or an anomalous pattern in decision confidence, signals that the agent's behavior has changed. Fingerprint anomaly alerts require no per-behavior threshold configuration: they fire when something changes, regardless of which specific dimension changed, making them valuable for detecting failure modes that no predefined metric captures.

Replay-as-test validates autonomous workflow behavior before deployment by using production trace sessions as test fixtures. Inputs, tool call sequences, and intermediate state captured from real production runs are replayed against a candidate agent configuration, with outputs compared to the production baseline using exact-match and evaluator-as-judge scoring. The replay harness must use deterministic tool-call stubs that return recorded production responses. Integrating the harness into the CI pipeline as a required gate before promotion makes it a structural quality control mechanism.

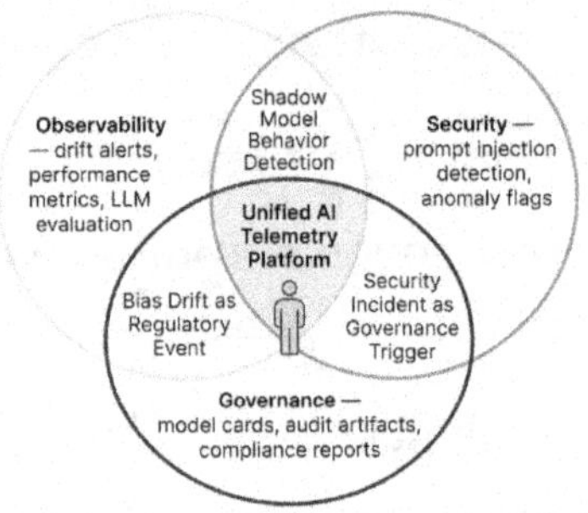

A unified platform reduces redundant pipelines and yields richer correlated signals.

Behavioral fingerprint similarity matrix across seven agent types, showing normal intra-agent clustering and two anomalous cross-agent sessions identified by fingerprint-based correlation

16.4 Closed-Loop Control: From Observability Signal to Autonomous Remediation

Closed-loop control connects observability signals directly to control actions without requiring human intermediation for every response.

Observability-Driven Retraining and Prompt Tuning

Observability-driven retraining connects the drift detection layer to the training pipeline so that retraining initiates automatically when monitored metrics cross predefined thresholds. A stable design uses a holdout evaluation gate: the trigger fires only when PSI exceeds the threshold for a sustained window, and an offline evaluation of the retrained candidate against the golden dataset shows improvement over the current production model. The holdout gate prevents retraining noise from propagating into production.

Observability-driven prompt tuning applies the same pattern to LLM prompt templates. When evaluator-as-judge scores for a specific template version fall below the SLO threshold over a sustained window, the platform emits a prompt degradation event. The pipeline retrieves degraded sessions from the trace store, runs an optimization step using DSPy with degraded sessions as negative examples, and proposes a revised template validated against the golden evaluation dataset. If validation passes, the platform promotes the revised template to a canary fraction of traffic. Every template change must land in an immutable audit log. Prompt templates are effective code.

Edge AI Telemetry Pipeline

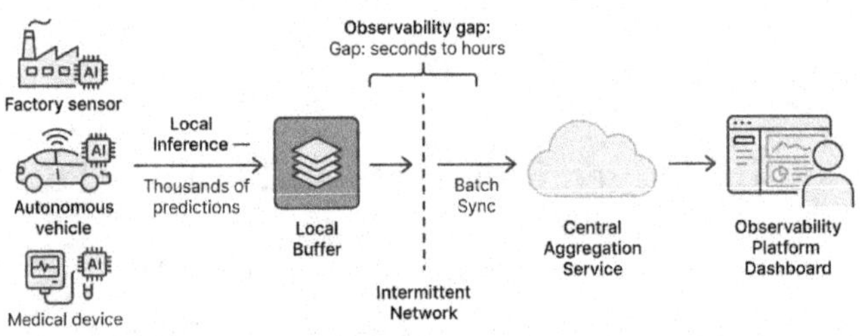

Edge AI observability requires explicit gap management and local fallback policies.

Closed-loop prompt tuning pipeline: evaluator-as-judge score stream feeding threshold policy registry, degraded session retrieval, DSPy optimization, holdout evaluation gate, and canary promotion with continuous score monitoring

Automated Rollback, Kill-Switch Propagation, and Eval-as-Spec

Automated rollback reverts a model, prompt template, or agent configuration to a previously validated artifact hash without human intervention when quality signals exceed rollback thresholds. Rollback policies require blast-radius awareness: a rollback affecting the primary model cascades to every dependent session. A graceful rollback drains in-flight sessions before switching; a hard rollback switches immediately. Kill-switch propagation is the coordinated suspension of work assignments across all instances of a specific agent type when quality signals cross a criticality threshold. The requirement is a kill-switch bus: a publish-subscribe channel the platform writes to with a signal specifying agent type identifier, tenant scope, and suspension mode. Every orchestrator and agent framework instance subscribes and enforces locally, with propagation latency targeted under five seconds. Each agent framework must expose a hook for kill-switch check logic; frameworks lacking it cannot participate in automated enforcement.

Eval-as-spec encodes the agent's behavioral specification as a machine-readable evaluation suite serving as the platform's authoritative definition of correct behavior. The suite defines permitted tool call sequences, output constraints, and input validity criteria. Failures are behavioral regressions. Online rule-based checks run within the session's latency budget; evaluator-as-judge scoring is deferred to an asynchronous batch pipeline.

Digital Twin Pre-Deployment Testing Flow

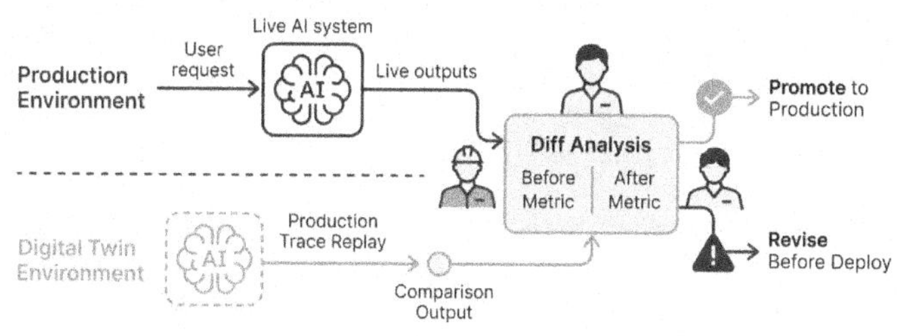

Shift-left observability catches regressions before they affect live users.

Kill-switch propagation architecture: observability platform emitting signal to pub-sub bus, orchestrator and framework instances enforcing suspension, and graceful drain of in-flight sessions before artifact rollback

The Platform-as-Agent Pattern and Formal Verification of Guardrails

The platform-as-agent pattern implements the observability platform as an agent with access to control tools, including the runbook library, the prompt template repository, the model serving control plane, the kill-switch bus, and the incident management system. An LLM-backed reasoning component generates a remediation plan based on the detected anomaly, the current system state, and historical incident context. The LLM component's output is always subject to a validation gate that checks the proposed action against permitted action types, the affected agent scope, and rollback reversibility before execution.

Formal verification of guardrails addresses the risk that automated control actions are themselves incorrect or harmful. Guardrail constraints are formally specified, machine-readable, and enforced at the control action execution layer: no suspension of more than 30 percent of active agent capacity in a 60-second window without human confirmation; no modification of a regulated model's prompt template without a governance-signed approval token.

Property-based testing validates guardrail enforcement logic under all reachable state transitions. The guardrail layer is safety-critical: a bug that incorrectly suspends a critical workflow can cause an outage more severe than the regression it was intended to prevent.

Open Standards AI Instrumentation Stack

Adopting open standards at the data layer prevents costly vendor lock-in.

Platform-as-agent control loop: monitoring pipeline detecting policy violations, LLM reasoning component generating remediation plan, guardrail enforcement gate validating scope and reversibility, and audit event emission to an immutable log

16.5 Open Problems and Engineering Bets

This section addresses problems in which the engineering path is only partially clear. Still, the full solution does not yet exist, and the technology bets platform teams should make now to position themselves as the field matures.

Continual Learning Monitoring and On-Device Edge Observability

Continual learning can lead to catastrophic forgetting: incremental model updates improve performance on new data while degrading performance on previously learned patterns. Detecting it requires per-slice accuracy tracking over the full historical input distribution, not just a rolling window. No major platform vendor has solved this out of the box. Teams running continual learning workloads should plan to build this capability custom.

On-device edge inference observability monitors models running on intermittently connected hardware outside the data center. The architecture buffers telemetry locally, synchronizes with the central store upon connectivity restoration, and then reconciles batched telemetry. Open problems include local anomaly detection within the device's compute and power budgets, causal ordering reconstruction when batches from multiple devices arrive out of order, and timestamp correction to account for clock drift across the connectivity gap.

AI Observability as Strategic Organizational Capability

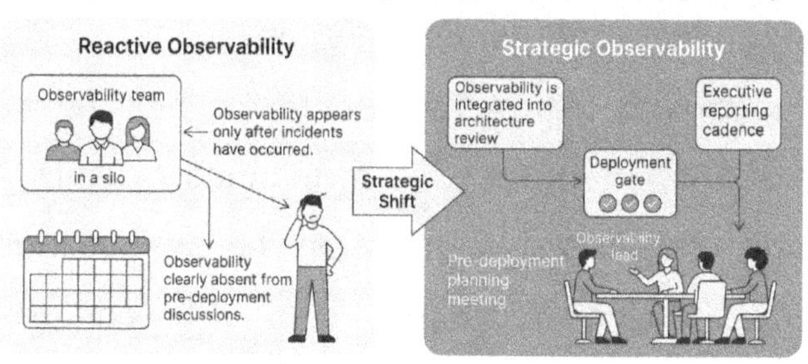

Observability becomes strategic when it precedes deployment decisions, not follows incidents.

Edge inference observability architecture: lightweight OTel agent on constrained device, local buffer with overflow policy, batch synchronization on connectivity restore, and clock-drift correction at central reconciliation

Privacy-Preserving Telemetry, Explainability at Scale, and AIOps Recursion

Differential privacy telemetry enables population-level monitoring of AI systems that process sensitive personal data without retaining raw inference inputs. A practical two-tier architecture: a differentially private aggregate tier with indefinite retention feeds population-level anomaly detection; a short-retention raw sample tier under strict access controls feeds session-level investigation. Explainability tooling at scale remains an open problem for multi-agent pipelines. The span tree provides a structural decomposition of a pipeline's execution, but not feature-level attribution for the final output. Teams should capture full-span trees and session-level audit logs as raw material for whatever explainability tooling emerges.

AIOps recursion requires a monitoring hierarchy where the observability platform's own telemetry feeds into a deterministic, rule-based foundation layer that does not depend on AI-based reasoning. The AI-based reasoning component operates on top of that foundation and is subject to the same quality and availability SLOs as any other production AI system.

Pilot-to-Production Observability Gate for New AI Paradigms

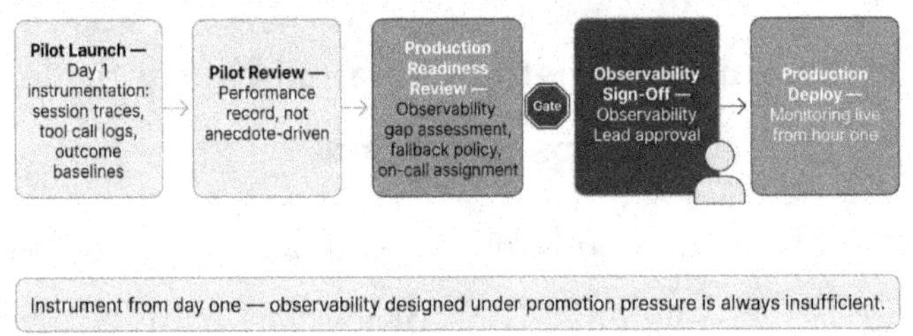

Two-tier privacy-preserving telemetry architecture: differential privacy aggregate tier with indefinite retention feeding population-level anomaly detection, and short-retention raw sample tier under access control feeding session-level investigation

Standards Landscape and the Build-Versus-Wait Decision

OpenTelemetry GenAI semantic conventions define attribute schemas for spans emitted by LLM API calls, embedding model invocations, vector database queries, and agent tool calls. LLM span conventions are approaching a stable state; agent-to-agent communication conventions are under active development. OpenInference, developed initially by Arize and adopted by a growing number of vendors, defines a complementary schema optimized for LLM tracing and evaluation. The practical guidance is to instrument against OpenTelemetry GenAI for all new deployments and emit OpenInference attributes as a supplemental layer where the backend requires them.

For the build-versus-wait decision: LLM span instrumentation and basic agentic trace collection are mature enough to adopt from open-source implementations. For cross-agent federated trace assembly, behavioral fingerprinting, and the platform-as-agent control loop, the tooling is at the prototype stage; we plan to build custom solutions with vendor offerings that mature over two to three years. For continual learning, monitoring, and differential privacy telemetry, build custom or partner with vendors. The instrumentation layer is the most expensive component to replace, and getting it right, vendor-neutral and standards-aligned, before the platform is at scale, is the single highest-return investment in the observability program.

Four Commitments
Three-Year AI Observability Mandate

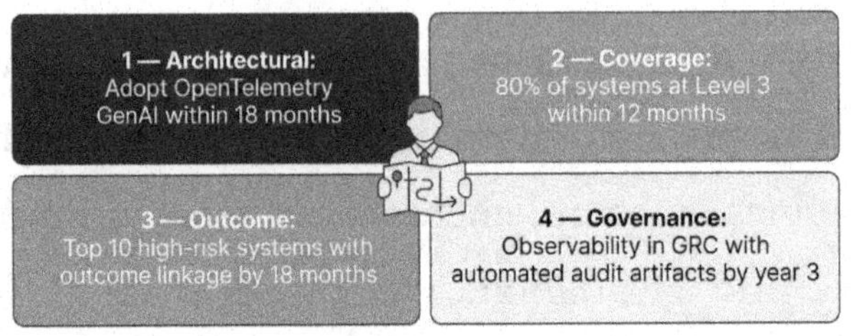

Four concrete commitments convert AI observability ambition into measurable program outcomes.

Standards alignment roadmap: OpenTelemetry GenAI stable LLM spans as the production baseline, OpenInference as a supplemental export layer, and active development areas for agent-to-agent conventions and multimodal input schemas

16.6 Technical Team Checklist: Forward Investment Bets

- **Action:** Instrument every agent framework with W3C traceparent propagation across all inter-agent HTTP calls and message queue handoffs using OpenTelemetry messaging instrumentation libraries. **Owner:** ML Platform Eng. **Done when:** A complete parent-child span tree is reconstructable for any multi-agent session from a trace store query alone, with no missing root spans from context propagation failures.
- **Action:** Implement a behavioral fingerprinting pipeline computing per-agent feature vectors (tool call sequence

distribution, inter-tool latency distribution, output token count distribution) over rolling session windows with cosine similarity anomaly detection against a 30-day baseline. **Owner:** ML Platform Eng. **Done when:** Fingerprint anomaly alerts fire within 15 minutes of a behavior shift, with a false-positive rate under 10 percent over a 90-day evaluation period.

- **Action:** Deploy a replay-as-test harness replaying a corpus of production agent sessions against candidate configuration changes using deterministic tool stubs, integrated into the CI pipeline as a required gate before promotion. **Owner:** ML Engineer. **Done when:** Every agent configuration change is validated against at least 200 production session replays before any traffic reaches the changed configuration.
- **Action:** Implement a kill-switch bus using a publish-subscribe messaging system and register kill-switch check hooks in all agent framework instances, with end-to-end propagation latency under five seconds from signal emission to enforcement. **Owner:** ML Platform Eng. **Done when:** A kill-switch test fire during a maintenance window demonstrates sub-five-second propagation across 100 percent of registered agent instances with confirmation events in the audit log.
- **Action:** Define eval-as-spec evaluation suites for every production agent type with programmatic constraint

checks and evaluator-as-judge assessments, run in streaming mode against the live trace feed. **Owner:** ML Engineer. **Done when:** Every production agent type has an eval suite with at least 10 behavioral constraints, and eval suite failures generate on-call alerts within 2 minutes.

- **Action:** Specify guardrail policies for the platform's automated control actions, including capacity suspension limits and regulated-model change approval requirements, and implement property-based tests validating guardrail enforcement logic under all reachable state transitions. **Owner:** Architect. **Done when:** All guardrail policies pass property-based test suites with at least 10,000 state transitions per policy.
- **Action:** Design a two-tier telemetry architecture with a differentially private aggregate tier and a short-retention raw sample tier for all agent workloads processing personal data, with privacy budget, sampling rate, and retention period documented in the governance artifact. **Owner:** Security Eng. **Done when:** All personal-data-processing agent deployments have a documented privacy-telemetry design, with an aggregate-tier anomaly-detection false-positive rate within 20 percent of the non-private baseline.
- **Action:** Adopt OpenTelemetry GenAI semantic conventions as the mandatory instrumentation standard

for all new agent deployments and publish an eighteen-month migration roadmap for existing deployments, with quarterly coverage percentage tracked as a platform SLO. **Owner:** ML Platform Eng. **Done when:** 100 percent of new deployments instrument against OpenTelemetry GenAI from day one, and portfolio migration coverage reaches 80 percent within eighteen months.

16.7 Team Conversation Starters

- When an autonomous agent session produces an incorrect outcome, how long does it take the platform team to reconstruct the full causal chain from initial task specification through every intermediate agent decision? What is the target MTTD, and which instrumentation gap is the largest contributor to the current baseline?
- If the platform's LLM-backed observability reasoning component silently degrades, which deterministic monitoring layer will detect it and page the on-call engineer? Has the monitoring hierarchy been designed so that the foundation layer does not depend on AI-based reasoning to detect AI reasoning failures?
- Which agent framework instances in production currently lack a kill-switch enforcement hook, and what is the blast-radius estimate for a quality regression in those agents given the absence of automated suspension capability?

- The eval-as-spec suites define what correct agent behavior looks like. Who owns those suites? How often are they reviewed against the current production distribution? And what is the process for updating a constraint when legitimate behavior has changed versus when the constraint itself was wrong?
- For agent workloads processing personal data, is the raw sample tier's access control and retention schedule documented and audited, or is raw personal inference data being retained in a trace store with broader access than the privacy policy allows?

16.8 Key Takeaway

The observability stack built for single-model inference endpoints will not scale to autonomous multi-agent systems, continuously-learning pipelines, or edge inference deployments outside the data center perimeter. The architecture that scales requires span linking and federated trace assembly to reconstruct causal graphs across organizational and tenant boundaries, behavioral fingerprinting as a probabilistic correlation and anomaly-detection mechanism, and closed-loop control that connects observability signals to remediation through kill-switch propagation, automated rollback, and the platform-as-agent pattern with formally verified guardrails. Eval-as-spec provides the formal behavioral specification mechanism.

OpenTelemetry GenAI semantic conventions and OpenInference provide the portable instrumentation foundation layer that survives vendor changes. Teams that build this architecture now will govern their autonomous systems with confidence. Teams that defer it will discover its absence during an incident.

17 Closing

The engineering thesis of this book can be stated plainly: instrumentation precedes governance, evaluation is first-class infrastructure, agents require traces, not aggregate metrics, and token cost is a runtime signal that belongs in the same pipeline as latency and error rate. Those are not philosophical positions. They are load-bearing design decisions that determine whether your production AI program survives real traffic, regulatory scrutiny, and the 3 AM incident that proves your monitoring gap was not hypothetical.

Every chapter has been building toward the same conclusion: an AI system you cannot observe is a system you cannot operate. A drift detector with no reference window cannot flag when a credit-scoring model has quietly shifted against a protected class. An LLM deployment with no output evaluation harness cannot detect when a prompt injection has redirected the model toward adversarial behavior. An agent loop with no span-linked trace cannot tell you which tool call triggered a cascade that burned thousands of dollars of API budget before the cost sentinel tripped. These are the predictable failure modes of under-instrumented systems, and every team that shipped AI without observability coverage has eventually discovered one of them the hard way.

Engineering the Path Forward

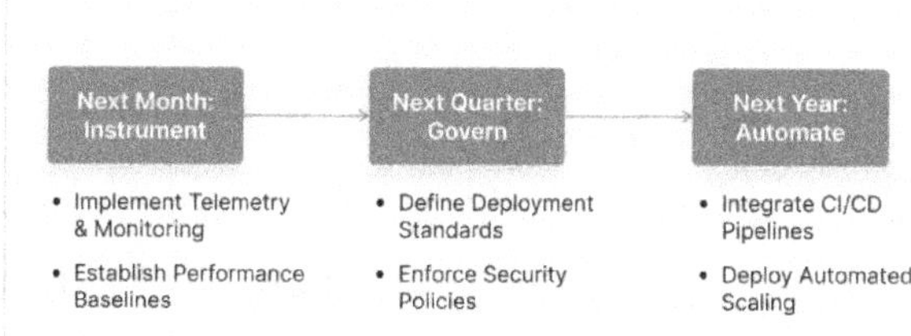

The five-layer observability stack from infrastructure health through agent decision tracing, showing the instrumentation path that converts raw telemetry into governance-ready audit artifacts

The five-layer stack, infrastructure health, data pipeline integrity, model performance, output evaluation, and agent decision tracing constitute the minimum viable telemetry surface for any system that a regulated enterprise must defend. Each layer feeds the next. GPU memory pressure and pod saturation are leading indicators of the latency spikes that corrupt output distributions. Data pipeline integrity metrics, schema conformance, null rates, and referential coverage are the prerequisites for trusting any feature-level drift signal. Output evaluation, whether via evaluator-as-judge scoring, a golden-dataset regression harness, or a human-review queue, catches semantic degradation that infrastructure metrics are blind to. Agent decision tracing, span-linked capture of every tool call, retrieval step, and branch in the agent loop, is the only data structure that supports meaningful post-incident attribution when an autonomous system performs an unintended action.

Ownership matters as much as architecture. The platform team owns the SLO for inference latency and the budget guardrail that gates agent tool-call spend. The ML engineer owns the eval harness and the cadence for golden dataset refreshes. The governance engineer owns the attestation pipeline that produces the audit log entries regulators will eventually request. The SRE owns the runbook that converts a drift alert at two in the morning into a decision path: roll back, restrict traffic via feature flag, escalate to the incident commander, or ride it out under elevated monitoring. These accountabilities must be wired into on-call rotations and team charters before the incident that makes the gaps visible.

Agentic AI will become the dominant enterprise deployment pattern within the next few years, and the observability baseline will shift accordingly. Full execution graph capture, every agent invocation, every inter-agent handoff, every tool call with its input and output, every retrieval query with its latency and result count, all linked by trace context that survives async boundaries and message queue hops, will move from advanced practice to table stakes. Teams building agent platforms without that trace infrastructure today are accumulating observability debt at the same rate that early microservices teams did before distributed tracing became an expectation. The debt is payable, but it compounds fastest during incidents.

Autonomous remediation will also mature. Platforms will initiate corrective actions, throttle a model endpoint via automated rollback, reweight a retrieval pipeline when cosine similarity distributions degrade, or isolate a misbehaving agent tool when its error rate crosses the kill-switch threshold. That automation requires governance infrastructure before it is enabled: immutable audit logs of every automated remediation action, human-in-the-loop approval gates for irreversible operations, and blast-radius bounds that limit autonomous scope without escalation.

The regulatory environment is converging toward explicit observability requirements faster than most engineering roadmaps are moving to meet it. The EU AI Act provisions on high-risk systems, the NIST AI RMF's emphasis on continuous monitoring, and SR 11-7 model risk guidance are not the ceiling. They are on the floor. Organizations with mature attestation pipelines, queryable audit logs with immutable write semantics, and documented model risk lineage will find that compliance is already built in. Organizations that have not built that infrastructure will retrofit it under external deadline pressure, which is the worst possible time to instrument a production system.

Commit to the following. Next month: instrument the highest-risk production model with output evaluation coverage over at least a sampled fraction of live traffic. Attach a cost sentinel to every agent deployment that enforces per-session spend limits and pages on-call when the daily aggregate spend crosses the budget threshold. Next quarter: stand up a shared observability platform that collapses infrastructure, model, and output telemetry into a single queryable backend with a unified alerting bus. Define SLOs for inference latency, drift rate, and output quality score for every production model, then wire those SLOs into error budgets that gate model update deployments. Next year: build the attestation pipeline that converts observability signals into governance artifacts. Design the agent trace schema to support cross-agent attribution across all multi-agent systems in the portfolio. Instrument token cost at the request level, route it into the FinOps pipeline, and establish unit-cost SLOs that make cost regression as visible as latency regression.

None of that work is optional at enterprise scale. The incident you have not had yet is not evidence that your monitoring is sufficient. It is evidence that the gap has not yet been exercised. Instrument now, before the next production AI failure reveals what was missing. The telemetry infrastructure you build before the incident is the difference between a post-incident report that reads "detected and contained in fourteen minutes" and one that reads "we are still determining the scope of impact." Build the first one.

www.ingramcontent.com/pod-product-compliance
Lightning Source LLC
LaVergne TN
LVHW010600100826
845148LV00014B/2780

* 9 7 9 8 9 0 4 9 8 0 0 7 8 *